International Disinformation

International Disinformation

A Handbook for Analysis and Response

By

Robert Kupiecki
Filip Bryjka
Tomasz Chłoń

BRILL

LEIDEN | BOSTON

Cover illustration: Disinformation media and abstract 3D illustration. © Alicja Nowakowska.

This book was prepared under the "Secure Automated Unified Framework for Exchange" (SAUFEX) project, financed by the European Union under the HORIZON EUROPE program.

Funded by the European Union

The Library of Congress Cataloging-in-Publication Data is available online at https://catalog.loc.gov
LC record available at https://lccn.loc.gov/2024045467

Typeface for the Latin, Greek, and Cyrillic scripts: "Brill". See and download: brill.com/brill-typeface.

ISBN 978-90-04-71575-2 (hardback)
ISBN 978-90-04-71576-9 (e-book)
DOI 10.1163/9789004715769

This book is printed on acid-free paper and produced in a sustainable manner.

Contents

PART 2
Disinformation: Recognition and Analysis

PART 3
Disinformation: Countering and Resilience

evolution of the entire information ecosystem, the technology boosting the news production and accelerating its circulation. These varied methodological approaches all contribute value within their individual disciplines but have left the study of disinformation disjointed and in need of a broader interdisciplinary framework – one that understands disinformation both as a social phenomenon and a tool of politics.

International disinformation, which is the focus of this book, adds to the foreign policy arsenals of modern states. Some states rely on it more heavily than others in their international activities, thereby becoming a paragon for other zealous states. However, what distinguishes this disinformation from other types of truth falsifications in human communication is, apart from its goals, the intensity and scale of means used to target wider groups. In a sense, state disinformers are like traveling salespeople trying to sell a certain message or narrative. However, unlike dishonest traders, they target not individuals but whole foreign societies, or large factions within them, seeking to control their behavior and decisions.

Disinformation, regardless of its motives and manifestations, poses a significant threat to security due to its increasingly harmful impact on communication and cooperation among the international community, both at the individual and interstate levels. True information is essential for individuals to satisfy their needs and function properly. The foundation of this process is trust, which is the recipient's belief that a source and its information are credible. The way information is perceived is sometimes conditioned by emotions, the appeal of the medium, or cognitive traps. It is therefore not difficult to imagine the potentially catastrophic consequences of automatic trust that is not supported by control mechanisms or sufficient critical thinking. It is essential that truthfulness of information be verified (McIntyre, 2018).

Contemporary technological developments, and in particular the rise of the internet and so-called new media, have resulted in an unprecedented production of information that circulates more quickly than ever before. Modern communication tools not only allow for access to various online services but also provide opportunities to store and develop knowledge, reduce communication costs, globalize messages, change social relations, and diversify forms of communication by combining words, pictures, sounds, and other content. As a result, modern opinion leaders are no longer solely entities legitimized by democratic choice (e.g., politicians) or public authority (e.g., ethical scientists or media journalists), but can also include any individual producer of disinformation.

These new technological tools also help to test the vulnerability of small niche interest groups and micro-target them more effectively as a result. This creates new opportunities to affect multiple social targets at the same time,

Introduction

Disinformation in its international form, often referred to as Foreign Information Manipulation and Interference (FIMI), is considered one of the oldest and most natural phenomena in social history. Throughout history, there have been numerous examples of communication techniques and processes that were built on lies, both spontaneously and in an organized manner (Frankfurt, 2005). And with advancements in technology and society, these communication techniques have become modern tools of politics and marketing (Phillips, 2019). Disinformation arises from the cognitive weaknesses, habits, and limitations inherent in the brain's natural physiology that controls information processing. These weaknesses are sometimes exploited consciously and deliberately by states in information operations to gain an advantage over the target of such actions.

An in-depth analysis of the international discussion on the phenomenon in question led the authors to propose an "operational" definition of disinformation, namely: *a doctrine and practice employed by states and non-state actors to deliberately use manipulated or falsified information in order to induce a desired change in a specific audience within a planned area of influence. Disinformation is intended to harm the recipients and is used as part of information and propaganda operations involving techniques of influence and psychological manipulation during times of peace, crisis, and war.* In Chapter 3, the authors outline various approaches to defining the phenomenon and explore alternative synonyms for conceptualizing it.

Disinformation is of interest as a subject of research in many scientific disciplines. Often, however, these disciplines are confined to methodological particularisms and a narrow spectrum of interests. For example, political scientists will look at disinformation as a tool of state and international policy; they will examine its goals and justifications and analyze the impact it has on society and individual and collective decisions. At the same time, a foreign relations scholar will study the place of this phenomenon in the foreign policy arsenals of modern countries. When addressing issues around the politicized impact of disinformation on democratic societies and their resilience, both academics may seek the support of a psychologist or sociologist in understanding the dynamics of social change and the limitations of cognitive processes. They may be assisted in this by a linguistic scholar focused on language and communication. Strategic studies researchers may view disinformation as a "peacetime and wartime weapon" and analyze the threats it poses. In the area of media studies, disinformation may be investigated through the lens of the evolution of the entire information ecosystem, the technology boosting the news production and accelerating its circulation. These varied methodological approaches all contribute value within their individual disciplines but have left the study of disinformation disjointed and in need of a broader interdisciplinary framework – one that understands disinformation both as a social phenomenon and a tool of politics.

International disinformation, which is the focus of this book, adds to the foreign policy arsenals of modern states. Some states rely on it more heavily than others in their international activities, thereby becoming a paragon for other zealous states. However, what distinguishes this disinformation from other types of truth falsifications in human communication is, apart from its goals, the intensity and scale of means used to target wider groups. In a sense, state disinformers are like traveling salespeople trying to sell a certain message or narrative. However, unlike dishonest traders, they target not individuals but whole foreign societies, or large factions within them, seeking to control their behavior and decisions.

Disinformation, regardless of its motives and manifestations, poses a significant threat to security due to its increasingly harmful impact on communication and cooperation among the international community, both at the individual and interstate levels. True information is essential for individuals to satisfy their needs and function properly. The foundation of this process is trust, which is the recipient's belief that a source and its information are credible. The way information is perceived is sometimes conditioned by emotions, the appeal of the medium, or cognitive traps. It is therefore not difficult to imagine the potentially catastrophic consequences of automatic trust that is not supported by control mechanisms or sufficient critical thinking. It is essential that truthfulness of information be verified (McIntyre, 2018).

Contemporary technological developments, and in particular the rise of the internet and so-called new media, have resulted in an unprecedented production of information that circulates more quickly than ever before. Modern communication tools not only allow for access to various online services but also provide opportunities to store and develop knowledge, reduce communication costs, globalize messages, change social relations, and diversify forms of communication by combining words, pictures, sounds, and other content. As a result, modern opinion leaders are no longer solely entities legitimized by democratic choice (e.g., politicians) or public authority (e.g., ethical scientists or media journalists), but can also include any individual producer of disinformation.

These new technological tools also help to test the vulnerability of small niche interest groups and micro-target them more effectively as a result. This creates new opportunities to affect multiple social targets at the same time, using various narratives within a strategic disinformation campaign implemented by a given state actor. In such operations, however, diverse content and ways of communicating that content – tailored to the specific nature of the target audience – also involve a shared intent and digital tools. Modern recipients of information, equipped with technologies, can independently choose what sources and information they want to access. However, these choices often contribute to the growth of "information bubbles" operating in competition. Such bubbles compete for the attention of recipients, striving to increase the "clickability" of their own content. To improve customer satisfaction, they will often simplify or falsify information to better correspond to their target demographics' identified or influenced needs, tastes, and emotions. In business, manipulating information translates itself into customer acquisition and, ultimately, financial profit. However, in interstate relations, manipulating information brings rewards of much greater value.

In this book, we look at the specific dimensions of this phenomenon, namely disinformation carried out by modern states and other international actors, sometimes used by states as proxies. From the earliest forms of inter-state relations, disinformation based on fabricated data has been a policy tool used by states in times of war and peace with the goal of gaining an advantage. It has been used to blur and disrupt opponents' recognition of the environment, therefore making it difficult to overcome uncertainty around decision-making.

In each case, these activities have specific justifications and contexts, but their general ambition seeks to improve the disinformer's position in the world and achieve specific short- or long-term benefits. Such benefits are often the opposite of cooperation and reflect a zero-sum game, measured by the resulting influence, power, and control the disinformers have over decisions and social processes in foreign countries. A well-known example of such consequences of disinformation is Russian influence over presidential campaigns in the U.S. and the weaponization of disinformation during the COVID-19 pandemic (Allcott & Gentzkow, 2017, pp. 211–236; Collins, 2021). Disinformers view information operations as a cheap tool of influence based on state power and facilitated by modern technology. However, the consequences of such operations are deep and destructive for democratic societies, whose resilience continues to be tested as means of counteraction remain relatively limited and often delayed.

This book addresses several research issues concerning three distinct groups of problems related to disinformation, which are:

1. The origins and characteristics of disinformation as a security problem for modern states.
2. Ways to identify disinformation and operational schemes of international disinformation actors.

3. Countering the threat of disinformation through methods that make societies more resilient and promote media education.

The community of democratic states is increasingly aware of disinformation threats, which has encouraged their societies, governments, international organizations, and technology companies to reflect on educational and protective measures. The key to successfully countering contemporary forms of disinformation is by making individuals and societies more resilient. Resilience is the ability to recognize and solve problems, assess a situation, and react to it appropriately. It also depends on the ability to respond to false, manipulated, or incorrectly prepared and disseminated information in a systematic manner.

The authors of this book took up the challenge of gathering in one place previously scattered knowledge on how to understand, recognize, and combat disinformation. We aim to offer readers a publication that, in a scientifically yet concise and reader-friendly manner, guides them through the meanders of this issue. To this end, the book has been divided into three parts, each consisting of four chapters and a short summary in the form of answers to the ten most relevant questions related to the topics discussed.

The first part of the book is devoted to the concept of disinformation, its intellectual and political origins, and its broader social context. The opening chapter discusses issues related to the evolution of the international security environment in recent years and how disinformation has developed into one of its principal threats. Chapter 1 also discusses the issues of the information ecosystem of modern countries, including the phenomena of post-truth and information science. Chapter 2 introduces the intellectual tradition related to the philosophical and legal dimensions of lies, as well as the foundations of strategic thought, which is the premise of the process known as the "weaponization of information". Chapter 3 contains a conceptual framework and a critical review of the definitions of disinformation in science and documents produced by states and international organizations. In this chapter, we also formulate our own working definition for the term. Chapter 4 describes the basic tools used in disinformation activities and their main development trends, covering the process of disinformation as a learning system; derivatives of the development of modern digital technologies like artificial intelligence; and the characteristics of China as a powerful state actor in the field of international disinformation that uses social media effectively and has a huge track-record of exploiting other media types for its propaganda and disinformation activities.

The second part of the book looks at recognizing and analyzing disinformation. Chapter 5 discusses the impact of disinformation on a state's information security in the theoretical, practical, and legal dimensions. In addition to a discussion of the general characteristics of information manipulation, its specialized forms, which include intelligence disinformation and military disinformation, are also presented. Chapter 6 identifies methods and techniques of disinformation as well as the goals pursued by political actors using this tool. The following section, Chapter 7, discusses methods and techniques of information analysis available to internet users, including critical thinking, fact-checking, and open-source intelligence. This chapter contains a set of useful tools that can be used to verify information and its sources. In Chapter 8, special attention is paid to the information and psychological operations utilized by Russia, which constitutes the foremost threat to the information security of NATO and European Union members.

The third part of the book covers issues related to counteracting disinformation and building resilience, ranging from the roles of individual participants in the information community to those of state authorities and international organizations. In this context, chapter 9 then offers a general characteristic of the challenges facing modern democratic societies. Chapter 10 points to the role of media education as a key instrument for counteracting disinformation and reflects comprehensively on national models and practices as well as relevant recommendations of the European Union. It also contains an overview of methods of combating disinformation at the level of individual internet users. Chapter 11 outlines the activities of tech companies and social media platforms that propel the circulation of information, such as Facebook, Twitter and YouTube. It analyzes how they react to omnipresent misinformation, offers recommendations on what more could be done to reduce it, and evaluates the expectations of users, regulators, and research and journalism communities towards these companies. It also examines the role of civil society in counteracting disinformation. The last section of the book, Chapter 12, is devoted to the responses of states and the international community, primarily the European Union and NATO.

We intend for this book to be both a contribution to and strong proponent for interdisciplinary research on disinformation. However, it is itself embedded in research methods and areas of interest in strategic and security studies. The bibliography contains a variety of source texts, monographs, scientific articles, and analytical materials produced by international organizations.

The main part of this book was written before Russia's war of aggression against Ukraine began on February 24, 2022. When possible, we have endeavored to analyze related issues.

The research and writing of this book have been greatly facilitated by a generous scientific grant offered by the EU's Horizon Europe program (HORIZON-CL2–2023-DEMOCRACY-01–01, Secure Automated Unified Framework for Exchange – SAUFEX).

PART 1

Disinformation: Environment, Concept, and Threats

CHAPTER 1

The Security Environment and Information Ecosystem of Modern States

The classic approach to security as defense of a state territory against military aggression and national survival invariably lies at the center of analysis of politics and international relations. However, since the end of World War II and the advent of the nuclear and information age, this concept has been systematically evolving, as has the corresponding security policy paradigm.

1 Disinformation and the Security Environment of Modern States

The growing complexity of the international environment affects an increasing number and diversity of participants and the dynamics of cooperation and rivalry they produce. The patterns of their mutual interactions in the globalized world have also called into question the classic divisions of domestic and international security. This raises not only difficulties in recognizing the origin, scope, and consequences of a given problem but also the emergence of new categories. Many of these issues operate, as it were, between and at the interface of the above-mentioned classifications. Those security issues arising within states often are not limited to interactions only within their interior, and those of a transnational nature may create local and regional mutations. Within this intermestic framework, global scope and local specificity, conditioned by changing circumstances or the persistence of cultural contexts, as well as the challenges posed by disinformation, should also be considered.

Three fundamental factors in recent decades have accelerated the emergence of the modern security paradigm:

1. The progressive evolution of the military factor, which traditionally monopolized thinking about state security and focused on analysis of international security. It is characterized, on the one hand, by demonopolization and, as a result, a new, more inclusive security paradigm. This evolution extends the spectrum of issues and corresponding means used by national, non-state, and supra-state entities to pursue security policies. On the other hand, in this process, a paradox exists, which recognizes a widening re-militarization due to the ongoing "weaponization" of inherently non-military means.

DOI:10.1163/9789004715769_003

2. The delegitimization (though not the disappearance, as evidenced by Russia's undeclared war with Ukraine) of classic interstate wars as a conflict resolution tool. This is due to, on the one hand, the development and dissemination of the principles of international law and the intensification of interstate ties. They make it difficult to isolate a given conflict from the interests of the wider international community. On the other hand, this development is also the result of nuclear weapons and the mechanism of *mutually assured destruction* (MAD), which has brought stability to relations between superpowers. Concern around global annihilation quickly blurred the line between *pyrrhic* political and military success in thermonuclear conflict and unimaginable failure. In these conditions, the essence of security strategy had to focus on prevention of war in any form as it carried the risk of uncontrolled escalation. As was eloquently put by Henry Kissinger, the purpose of war cannot be a war victory in its strict sense but the achievement of specific political conditions fully understood by the enemy. "Our purpose is to affect the will of the enemy, not to destroy him" (Kissinger, 1969, p. 189). To do so now requires deterrence involving a combination of information and propaganda efforts as well as military, economic, social, informational, educational, political, and diplomatic measures.
3. A change in the understanding of security marked by successive "revolutions in military affairs". These influenced thinking about conventional conflicts, regarding both their forms (from limited wars to political ones conducted with hybrid methods in the "gray zone between war and peace") and operational aspects, including precision and autonomous weapons, automated command, computers, information domination, psychological operations, decentralization, and asymmetry of actions. No conflict is bound by a single-dimensional space, nor does it play out solely at the military level. It does not end when victory is declared by one side, nor is it limited to the duration of the fighting. In this new security paradigm, information plays an important role at all stages.

In 1966, future Nobel Prize winner Thomas Schelling wrote that "the concept of victory inadequately describes what society today expects from its military: deterrence resulting from its existence, an instrument strengthening the power of negotiation as a derivative of the acquired ability to kill, and not the effect of victorious fights…" (Schelling, 1966, p. 31). This modern understanding of the costs of war involving superpowers does not directly put the military factor in the foreground of their mutual rivalry. However, it invariably remains the "last resort" and a component of deterrence strategy, reinforcing a message about readiness for war, creating offensive pressure, and guaranteeing a potential escalation of the conflict. Such logic can be found, for example, in the

Russian military actions and provocations in 2021 and 2022 on the border with Ukraine, which was accompanied by its aggressive demands towards NATO and the West. In February 2022, these demands then escalated into Moscow's open aggression against this country. It was accompanied by hybrid scenarios and indirect actions using, for example, surrogate actors (Bryjka, 2021a) as Moscow's systemic disinformation turned against the West.

The concept of security, and thus international politics, is today much broader than its military connotations. In the interconnected world, predictability, market stability, access to raw materials, technological development, information security, and human security issues continue to broaden the scope of national policies and the international agenda (Baldwin, 1997). Sovereign states that control their physical borders also face the challenge of securing the sovereignty of cyberspace – their virtual borders – where many economic activities and tools for managing critical infrastructure components and public policy decisions are transferred. The protection of the infosphere influences social communication and relations between social groups. Due to its openness, liberal rules of organization, the speed at which data circulates through modern information technologies, and the spontaneity of the content produced, the infosphere has become more sensitive to hacker attacks and hostile disinformation influences, which has led to the creation of an *alternative reality* around manipulated facts and their interpretations.

The security paradigm has been irreversibly changed by the blurring of boundaries between internal and international security; the transnational nature of threats to health, climate, and development; and hybrid scenarios driven by disinformation. The transformation has not been linear and unidirectional, however. It is possible to observe the evolution of institutions in their space and the growing normative void. They have three characteristics.

First, in certain segments of international security, the legal foundations and mechanisms of multilateral cooperation have undergone decomposition, particularly in the realm of conventional arms control and confidence-building measures. Second, problems in international security also concern normative and operational aspects. In the area of non-proliferation of weapons of mass destruction, for example, systemic foundations exist, but they are bypassed by some states and weakened or poorly implemented by others. Third, the slow-moving nature of normative and regulatory responses on the part of the international community remains problematic, particularly in new areas considered important for security like cyberspace, disinformation, and artificial intelligence and its derivative autonomous systems.

In the face of these dynamics, the concept of security has *de facto* become an open structure formulated in the discourse between the political factor and public opinion. As part of this process, securitization is progressing, understood by researchers of the Copenhagen School as an act of speech construction and not simply a reflection of social reality (Buzan et al., 1998). Such a mechanism itself is conducive to deliberate manipulation. Balzacq (2005) has further proposed that the securitization mechanism should be perceived as a correlation of three factors: a specific group of recipients, the social context, and the involvement of the political factor.

In public discourses, this strengthens the tendency to designate a broad definition of security threats and empower non-state actors in this framework. Disregarding threats to security is, of course, an unacceptable attitude for researchers and government leaders, but conceptual exaggeration also makes their work considerably more challenging. Tools of political and armed struggle, such as terrorism and disinformation, are nowadays treated not only as a tactic of asymmetric conflicts and an activator for various hybrid scenarios but also as separate threats with their own characteristics that demand specific preventive measures and defense (Ball, 2018). This form of securitization and its conceptual chaos, which broadens the operational field and geographic horizon of policies, has nevertheless de-monopolized the military factor in pursuit of national security. It has also emphasized the value of predicting and preventing broadly defined crises, solving them at the earliest possible stage, and analyzing the components of their variable scenarios.

Disinformation also presents a challenge as the importance of information in the modern world allows it to transform "ordinary" matters into security challenges. The development of the internet and social media has changed the architecture of social communication and redefined the power centers that shape public opinion. It has also become more difficult to protect the security of information and its processing in the communication space of a given country as state actors struggle to combat the rising influence of foreign disinformation. In this way, modern states have also received a powerful tool for directly influencing foreign societies as well as their policies and decision-making processes.

In many crises, be they political, economic, military, internal, or international, information operations constitute a recurring feature as a preparatory factor for the use of other measures. They are also used as a means of directly influencing the opposing party's perceptions and political behavior. When they are of a hostile nature and are intended to deliberately harm recipients, false or falsified information used for this purpose becomes disinformation. This distinction is important insofar as false information is a deliberately prepared message with the intent to harm the recipient. On the other hand, falsified information refers to various types of propaganda and manipulations, such as the mixing of falsehood and truth or a half-truth given in a false context (Lasswell, 2013; Lasswell et al., 1939; Lasswell et al. 1946; Laswell et al., 1980).

As early as 1970, a Canadian media researcher predicted that a future war would involve "party-based information activities, with no distinction between civilian and military participation" (McLuhan, 1970, p. 66). Much has been written on this subject, analyzing Russian information operations during the 2016 and 2020 presidential elections in the U.S., external influences preceding the British referendum on leaving the European Union, and Chinese propaganda regarding responsibility for the spread of COVID-19 (Kupiecki et al., 2021, pp. 33–70).

In the contemporary security environment, information is one of the most important strategic resources, both for political conflict and strengthening inter-state relations and domestic social order. As a result, its protection has become a priority (Aleksandrowicz, 2016, pp. 110–128). However, experts disagree as to the actual effectiveness of disinformation in the Western world. They continue to debate whether disinformation activities "parasite" the disabilities of cognitive processes and existing social divisions within democratic societies, or whether these activities have the ability to actually create them. Aleksander Lanoszka (2019) claims that "in an anarchic world, disinformation cannot be effective, because states do not trust sources located abroad, and the division of opinions among individuals makes it difficult for disinformers to focus their activities on sufficiently large homogeneous target groups" (p. 234). Gerrits (2018), speaking on the effectiveness of disinformation, contends that it is at best a "soft security" challenge (p. 21). These contentions, however, do not change the fact that policies based on the ability to anticipate threats are better than those that ignore them or are unable to recognize them.

At the same time, nation-states increasingly see their limitations in dealing with this new transnational architecture of security problems. This also applies to superpowers, which have greater means of preventing and combating such phenomena on their own. In their actions, however, they prefer the division of responsibilities and costs resulting from international cooperation. This is true at least in the declarative aspect of international security management. In practice, superpowers are both a desired participant in global solutions and a source of problems. The systemic disinformation practiced by Russia and China is a perfect illustration of this.

Since 2014, there has been a significant increase in online offensive by entities that are part of Russian state structures or act on their behalf and conduct mass-scale disinformation campaigns aimed at NATO and EU countries. This phenomenon, however, is not limited to Russia alone. Similar activities have also been undertaken by other participants of international relations who aspire to have global or regional superpower roles. The cases of China

(Karásková, 2020) and Iran (Kasapoglu, 2020) are well described in this regard. It is important to note that all states use information influence tools and modern communication techniques; however, not all of them use systemic disinformation or prepare information strings, nor do all of them make disinformation a permanent instrument of politics, diplomacy, and offensive influence.

Disinformation is also not only the domain of state actors. Activities in this area are also carried out by non-state entities, including terrorist organizations like Daesh (Winter, 2019). Given the unconventional nature of this form of hostile action, its essence remains the same as that of states applying tools of traditional conflict according to a "Clausewitzian logic" of warfare. Its focus is on effectively neutralizing all sources of an opponent's power and breaking their will to resist, gaining a mental or physical advantage, and controlling their assessment of the situation. It further aims to weaken the enemies' ability to make the right decisions, and, as a result, force them to recognize their own defeat. The "five circles" concept developed in 1995 by John Warden clarifies the spectrum of such activities. It depicts an understanding of the enemy's vulnerable points as five overlapping rings formed (from the inside) by leadership, system foundations, infrastructure, population, and its army (Warden, 1995).

Disinformation is a *sui generis* threat to modern societies. It disturbs the relations between components of the information ecosystem. As a result, it influences individual and group perceptions of reality by distorting data. It may also contribute to limiting the space for accepting political decisions in democratic countries. This mechanism (admittedly in isolation from the issue of disinformation) was described in the 1990s as "Overton's window". This concept touches upon the social determinants of a change in collective attitudes towards politics, which are also used by information operations operators (*The Overton Window,* 2009; Astor, 2019). And although it can be relatively easy to detect the existence of disinformation and even deduce the nature of its strategic use by enemy states, assessing its effectiveness is quite complex. It can be highly impacted by the time, place, and context of the observation, the level of social resilience and media education of possible disinformation target populations.

However, distorted information intentionally used against someone may be a factor that, combined with other measures, creates a threat scenario described as *hybrid*. Hybrid threats represent a huge spectrum of security problems, which are both ambiguous and essentially unlimited in possible iterations, making them difficult to counteract. This characterization also seems to define the nature of threats to the democratic West today. These types of threats take advantage of liberal institutions and public trust, which are perceived as a weakness by determined entities that do not intend to obey the democratic rules of the political game. The hybrid nature of contemporary threats (Najzer, 2020) refers to actors using this type of method in various domains of potential conflicts, including the economy, culture, defense, and cyberspace.

According to experts from the team of analysts led by Gregory Treverton (2018), six factors justify the serious treatment of hybrid threats as a component of the modern security environment. These are:

1. Deregulation of the liberal international order and a shift in the definition of power from its material factors to the ability to influence and effect change over other states' perception, decisions, and ability to act effectively. This also means a redefinition of the rules governing international relations by revisionist states seeking to improve their own position.
2. The effects of globalization, expressed in increased networks of connections, which enhance the effectiveness of actions by smaller states and non-state actors that deprive states of a real monopoly on international operations.
3. The development of modern technologies, expanding the field of competition into cyberspace, which largely falls outside the regulations of international law and challenges traditional strategic thinking subordinated to the rules of territoriality.
4. Changes in the information space resulting from the development of the internet and social media. These contribute to the acceleration of the circulation of information but also increase the risk of a different understanding of it in different places around the world.
5. A change in the character of modern conflicts by reducing the role of the military factor and the act of physical struggle within conflicts. Conflict understood as a non-kinetic clash, primarily of societies rather than armies, blurs the differences between the state of war and peace and feeds hybrid scenarios.
6. Generational and cultural changes, shifting the emphasis on the determinants of security policies from the rivalry of superpowers and the risk of nuclear holocaust to the challenges of globalization, emphasizing interdependence as a new challenge for historical memory, identity, and the behavior of individuals and nations (Treverton et al., 2018, pp. 1–2).

Disinformation, which influences societies, may prepare a favorable environment for scenarios of hybrid actions as well as precede or accompany armed aggression, masking its intentions and justifying contradictions to the spirit of international law. In times of peace, disinformation may be a simpler and more effective way of influencing opponents – one that is cheaper and less controversial than traditional political, economic, or military means of coercion. Acting openly in the information space of a foreign state through secret services constitutes an act of diversion or espionage, punishable by law within the injured state. Similar activity can be masked by the activities of lobbyists masquerading as engaged citizens or civil society organizations, which are housed in a wide space of freedom of speech and legally protected within the democratic world. Their actions can arouse anxiety, maintain chaos, distract public attention, influence the results of elections in foreign countries, spread distrust of the truth, deepen divisions, and permanently change social attitudes towards various issues (Jackson, 2018). The cumulative effect of these influences in the hands of revisionist powers has become a tool for undermining the liberal international order and influencing social behavior abroad (Theohary, 2018).

Disinformation poses a significant threat not only to public trust and democratic processes but also to the global economy. A study by Canadian-German consultancy firm Prevency estimates that annual losses due to disinformation amount to USD 78 billion. These losses can include image damage (USD 17 billion), a decline in stock value (USD 39 billion), long-term harm to reputation (USD 10 billion), and costs for prevention campaigns. The study also found that one in four companies have been affected by disinformation and 74% of CEOs of large companies consider it to be the main cyber threat facing their businesses (*What is disinformation?*, 2018). Cybercrime is an ally of disinformation, and illegal knowledge about people and state affairs obtained in this way, when properly publicized, is capable of wreaking havoc on the fragile democratic consensus of Western societies. Disinformation can exacerbate societal divisions, fuel emotions, and alter public attitudes. Efforts to counter disinformation by authorities or affected parties can also lead to further confusion and amplification of *information noise* that deepens social confusion.

The information space has therefore become a place of international competition. Unlike democratic states which pursue propaganda but do not base their policies and international communication on systematic use of disinformation, Russia and China (which has been quick to scale up its efforts) include information activities as integral components of their foreign policy. Thus, the boundaries between disinformation in the service of the state's military interests during war and peacetime propaganda have become blurred. In many cases, these activities are conducted in a coordinated manner, directing the blade of their influence operations against the U.S., Western nations, and their organizations – namely, NATO and the European Union.

2 Post-Truth: Harmful Equality between Truth and Falsehood

The modern world is filled with information available on demand in immeasurable amounts, with the technical tools needed to access it possessed by most people. This new information environment is often referred to as a *post-truth* or post-fact world, in which the importance of objective truth has decreased or is completely indistinguishable from falsehood (Cooke, 2018). In an extremely cynical form, according to a prominent Kremlin propagandist, the objectivity of information "is a myth that is proposed and imposed on us" (Yaffa, 2014). As Francis Fukuyama claims, in the post-fact world, "essentially all authoritative information and its sources are confronted with contradicting facts of dubious quality and origin. In a world without fuses, there is no reason to believe that true information will win over false" (Fukuyama, 2017). In a post-fact world, therefore, truth and falsehood can co-exist on equal footing for recipients who are guided by personal convictions and emotions rather than a desire to confirm the facts. McIntyre (2018) defines this as "the recognition of an ideological advantage through which practitioners try to force someone to believe something, irrespective of existing evidence". Researchers consider the phenomenon of post-truth to be evidence of a triple revolution that is: technological, which affects the means of transmitting information; social, which accounts for cognitive processes and the effects of information processing; and anthropological, or the functioning of an individual within its community (Ferraris, 2019).

With the excess of information available, people look for safety by selecting opinions and explanations that confirm their previous views and experiences. Even if they change their minds, the process can occur without violating the rules of the "information bubble" in which they exist. This phenomenon has been accompanied by the transformation of the information space, where traditional sources of information are displaced by social media. These new sources produce and disseminate a plethora of unverified data and defend such carelessness as acts of freedom. Currently, information can be published by anyone – a professional journalist, a citizen convinced of their mission, a blogger, or a social media influencer. Internet users are not only recipients of messages; they also have a direct influence on the creation and dissemination of messages by sharing or commenting on the content of their choosing.

However, this shifting information environment is generally poorly reflected through adaptation of the educational system, which fails to prepare people to cope with this problem. In Europe, for instance, people are generally not taught to verify information from the internet, a trend that is particularly prevalent within older generations. The consequence is a society that increasingly functions in a gray zone between facts, opinions, and fabricated reality. Furthermore, social polarization and demographic changes deepen differences in opinions and undermine trust in sources of information that do not support the views of a given social group (Kavanagh & Rich, 2018, pp. 79–190).

Consequently, the same researchers point to four related trends supporting this phenomenon:

- A lack of agreement as to the facts and their analytical interpretation, often despite existing evidence.
- A blurring of the boundaries between facts and opinions in a way that makes it impossible to distinguish them.
- A growing number and domination of personal interpretations and opinions over facts.
- A diminishing trust in traditionally respected sources of information and facts (Kavanagh & Rich, 2018, pp. 21–40).

These trends have led to the dominance of disinforming opinions and personal positions in the public discourse over credible information and its objectified analysis. This lowers the quality of public debate and infects other areas of social trust with subjectivism and exclusion, including, for example, science or authorities. The consequence of this is the extreme individualization of decision-making processes concerning, for instance, health or investments, in which the role of knowledge and messages from public trust institutions is diminishing. To maintain support or neutralize a potential protest, these institutions may choose to legitimize views and attitudes built on disinformation. The erosion of public debate in the face of a scarcity of "common facts" and mutual trust leads to the alienation of social groups, as divergent perceptions of reality make it difficult to come to a collective consensus. The consequence of this is sometimes political paralysis resulting from the inability to reach an agreement and compromise. This, in turn, becomes a burden on all areas of the state's activity, also transferring to the level of international cooperation. Under such conditions, distrust towards state institutions and the extensions of international organizations continues to grow (Kavanagh & Rich, 2018, pp. 191–222).

The modern world could, therefore, be more aptly designated as the *disinformation era*. Its universality, associated with technological advancements of the media and supported by Artificial Intelligence-enabled apps and neurobiological knowledge and the personalization of interactions, has taken on new forms (Levitin, 2016). They range from tailored commercial ad prompts and news suggestions, based on machine analysis of one's online activity, through individual electoral preferences profiling, to new chatbots skills interacting with human needs.

The notion of information warfare, or information and psychological operations, closely embedded in the political strategies of states in times of war and peace, is one of the key defining characteristics of the disinformation era. The goal of disinformers acting in the service of their politics is to take control of their opponent's communication space, which is founded on knowledge, science, common sense, and democratic values, and disseminate their message within it to weaken social cohesion. This dissemination of intentionally crafted information gives the disinformer the chance to then shape the attitudes of recipients.

The credibility of scientific research and the opinion of experts or authorities in a given field are commonly probed by bloggers, vloggers, celebrities, and self-proclaimed pundits, often anonymous or unknown by their specific achievements. Despite a lack of necessary knowledge, competencies, and qualifications, these actors gain popularity on social media and garner captive audiences of hundreds of thousands of followers on Facebook, Instagram, YouTube, and Twitter, imparting a greater impact on society than reliable and credible institutions. Often, these well-known social media personalities are paid participants in promotional campaigns (Schuman, 2020).

One natural ally of disinformers is human stupidity. The term "stupidity" is used intentionally, understanding its unscientific and difficult-to-define nature. In this case, the term is used to describe the consequences of a lack of knowledge or a refusal to know, which can lead to irrational behavior. This behavior can be accompanied by ignorance or carelessness, as well as a preference for the spontaneous creation and dissemination of deformed truth and half-truths as well as a selective use of facts, gossip, conspiracy theories, or careless duplication of false content. It looks for content that is convincing and explains the world in simple terms, confirming the recipient's conviction. This can be delivered in the form of pseudo-knowledge supported by a statement derived from an informal "authority" (like a celebrity), or an opinion based on scientific evidence. The role of "communication facilitators" may nowadays also be played by social media. They eagerly promote the activity of the self-born "cognitive authorities", which create communities of like-minded people around them. These communities are run by their own rules, rewarding participation with access to information, or by recognizing presence. On the other hand, they penalize mavericks and dissenters by banning them – denying access.

Disinformation also plays on attitudes that could be defined as an individual or collective tendency to self-deception. The difference here lies in the source of the legitimacy of certain content. In the case of an individual, they decide on their own about the directions of cognition, and in the case of a community, there is a tendency to imitate and self-affirm others' perception of matters. The latter is easier to defend and harder to oppose due to the group pressure that cultivates intolerance to different views. This leads the recipient to search for information that meets only narrow criteria of subjective truth, a biased interpretation of the facts, or anti-rational rationalization.

One example of this is Moscow's claims that the West betrayed Russia after the end of the Cold War and that NATO has deployed large numbers of troops on its border; hence, it argues, it has the right to claim its sphere of influence in Europe. It may also exploit or deliberately deepen false codes of memory, which are eagerly used in Russian disinformation campaigns (Legucka & Kupiecki, 2022).

Modern methods of marketing and advertising are based on the repetition of content and images, subliminal persuasion combining real and prepared data, and individualization of messages in online advertisements based on machine analysis of users' network behavior. The purpose of such activities is to target as precisely as possible and to catch the attention of recipients using information prepared especially for them. They are effectively a contemporary training ground for civilian information operations. State disinformation dispatchers observe the techniques and methods used and subsequently analyze their suitability for effectuating changes in customer preferences. The worlds of politics and global trade, though aiming to induce different actions, can be compared by their instrumentalization of information to achieve specific benefits.

Long-term interactions of this type can change attitudes and human decisions. To this day, there is a fierce dispute in the U.S. over the scope of Russian disinformation activities during the 2016 presidential election and their impact on Donald Trump's victory. Similarly, much controversy and interest of analysts has been aroused over the long-lasting campaign of supporters for Brexit and the support they received from abroad. Planned and deliberate disinformation therefore constitutes a special case, and the derivatives of its threats gain importance. Its aim is to permanently influence the behavior of large groups of people and to manipulate their attitudes. The tools used for this purpose are based on the systemic use of false information to destroy cultural or common-sense foundations for the functioning of societies, like trust in authorities, scientific knowledge, public institutions or established democratic procedures.

3 The Information Ecosystem of Modern States

The information ecosystem is a conventional and metaphorical concept that describes the information environment surrounding societies and its mutually interacting transmission and reception, physical, virtual, and cognitive components (Babik, 2014). Some researchers call for the study of information ecosystems at a localized level, which concerns the "people, practices, values, and technologies in specific local conditions" (Nardi & O'Day, 1999). The dynamics of these elements are of key importance for many social processes, including civic education and collective resistance to contemporary threats. Without access to sources of verified information and the possibility of exchanging opinions, it is difficult for individual and larger social groups to be described as free entities that are able to consciously make rational decisions. It is generally assumed that eight components, described below, characterize the modern information ecosystem.

3.1 *Information Needs of Recipients*

This relates to the topical, correct, and exhaustive nature of information and its adaptation to the interests and perceptive possibilities of the recipient. As Materska (2007) has noted, the modern function of information has changed significantly. Today, it does not so much reflect reality but is directed at controlling the behavior of individuals and social organizations (p. 42). It also satisfies consumer needs, like entertainment, and inspires activities in the communication space. The key issue in this respect is the credibility of the information, which refers to two elements: (a) the objective truthfulness of the message and (b) the recipient's recognition of the information as "sufficiently" true to guide their decisions. A striking paradox of this situation is the "equality of the ontological status" when information is objectively true and subjectively regarded as such.

3.2 *The Information Landscape*

The digitization of modern societies has broadened understandings of the information landscape beyond the horizon of traditional media and conventional communication. The mass adoption of social media and new technologies has created a new model of activity, resulting from the de-monopolization of information production. Information producers can be anyone who, regardless of intellectual qualifications, substantive preparation, and ethical reflection, believes that they should express their own opinion or attitude. The internet has contributed to a democratization (but also anarchization) in the field of access to and production of information. On the one hand, it has become a factor encouraging citizens to participate in public debate. On the other hand, it has increased threats to online privacy and multiplied the risk of misuse of information by offering new ways to influence the decisions of individuals and social groups.

3.3 *Dynamics of Information Production*

Information production is highly decentralized in the modern world, defined by many centers and sources of influence. The operation of information producers is facilitated by universal access to the internet and social media. It is also influenced by cultural changes, removing from many spontaneous participants of the infosphere a sense of responsibility for information introduced into circulation. This provides a space for both the commercialization of behaviors selling false, sensational, or scandalous content with high levels of "clickability" and increased temptation to use them as a tool for manipulation and disinformation by interested entities.

3.4 *Access to Information*

Access to information is determined by law and other regulatory instruments, available technological solutions, and the existing information transmission infrastructure. It is also associated with barriers to access that disrupt the flow and use of information. The latter may also be associated with the user's limitations resulting from education or IT exclusion, manipulation of the message content, the society's culture, language limitations, or the speed at which the information flows.

3.5 *Use of Information*

The use of information concerns how it is processed and interpreted by the sender and recipient, and how trusted it is.

3.6 *Influence of Information*

The influence of information results from its degree of acceptance by a recipient. Information can co-shape social reality; it is a carrier of knowledge about the world and a tool for shaping public awareness. Thus, it is an important tool in political communication at the national and international levels.

3.7 *Social Trust in Producers, Content, and Transmitters of Information*

Social trust is a source of support for rational, adequate, and timely decisions. The trust and credibility of information is the primary inspiration for a recipient to make rational and optimal decisions.

3.8 *Centers of Influence*

Centers of influence are people, organizations, and institutions responsible for shaping the content and information flows. Their place is determined by their political, economic, or professional status. However, due to the modern democratization of the information production process, it is now nearly impossible to fully control the credibility of content.

The world, understood as a "global information village", consists of diverse social networks connecting people and groups. The concept of information or *media literacy* has been the subject of research for half a century. In UNESCO documents, this skill is defined as "knowledge of the needs of information, identification, finding, evaluation, organization and effective use of information for solving personal, work-related or wider social problems" (UNESCO, 2003). Within this framework, educated recipients are aware of their information needs. They have sufficient skills to find the necessary data as well as select and process it. They are also able to properly read the context in which the message obtained by them functions and correctly interpret it against their initial information needs. In a perfect world, each piece of information would be credible and true, and the participants of a given network could make only rational and optimal decisions on this basis. Such a situation is described by UNESCO standards, which in its 2003 Prague Charter postulated the universal standard of information education as the key to the development and reduction of social differences, the guarantee of participation in the information society, and the increase of tolerance (UNESCO, 2003).

The real world, however, is far from meeting such standards; contemporary media and information platforms can equally become transmitters of distortions and disinformation. Information processed consciously by humans and understood traditionally is closed in words, images, sounds, and gestures. However, the modern understanding of its essence should be detached from people's will and consciousness, extending it to other objectively existing records based on technological achievements, such as electromagnetic impulse, computer bit, algorithms, or readings of various devices.

In broad terms, modern information should be considered content in any form that comes from the human environment, which can be processed, be understood, inspire thinking, shape attitudes, link with other data, and lead to meaningful action. Moreover, information can exist with limited distribution or completely without human will and consciousness. However, while living blissfully unaware, people and societies are exposed to the consequences of facts, including them being deliberately or unintentionally distorted.

The following two examples show the scale of the effects of the above-mentioned possible informational influences:

1. The modern person can correctly receive and interpret traffic lights on the streets, accurately read messages in the media, and logically link facts that guarantee appropriate decisions and relative safety in their everyday life. At the same time, their environment is constantly trying to influence their cognitive processes and decisions. This includes influencing what goods they purchase through marketing tactics as well as influencing political choices by shaping their interpretations of various events. In each of these cases, the line between objective truth, the "truth" recognized by the recipient, and the disinformation fed to them is not always evident to the recipient. Despite this, the decisions made as a result of these influences will differ, as will their effect.
2. The same person, as someone who is not a specialist, cannot interpret, for example, the results from a radar signal, which relies on correct reception and interpretation to help secure modern countries. However, history has seen cases of falsified data from early warning systems that have brought the world close to the brink of nuclear war. The consequences of deliberate attempts to distort the correctness of data from such systems by a foreign country or influence their interpretation by decision-makers could therefore have dire consequences. This is illustrated by a well-known event on September 26, 1983, when Soviet early warning radars detected an American nuclear attack. This system error, when correctly interpreted by Colonel Stanislav Petrov, prevented Moscow's counterattack and, quite possibly, a nuclear Armageddon.

Both examples lead to the same conclusion: there is a need for measures to protect the correctness of source information, the integrity of information processing systems, the verifiability of conclusions as the basis for decision-making processes, and the habit of critical thinking. Doing so, however, presents a greater challenge in the modern information environment than in the past. The new media has not only globalized and accelerated the circulation of information and multiplied its numbers but has also changed the hierarchy of information. This has opened a space for organized state disinformation, which feeds on emotions, superstitions, deficiencies in knowledge, and weak cognitive processes. The excess of data in circulation not only exploits but also perpetuates these processes.

4 Vulnerability of Cognitive Processes

Disinformation takes many forms, and its spread is facilitated by numerous human and systemic weaknesses. At the individual level, these vulnerabilities largely relate to the characteristics of cognitive processes, which is the structure of thinking, and the habits that accompany them. Various types of "cognitive filters" imposed on information processing additionally open the door to increased vulnerability to disinformation. Authority, mentality, and emotions push people to explore the world through trust and familiar categories and subconsciously exclude matters that elude previous experience, particularly if they contradict well-established knowledge and beliefs. This is known as "authority and mental filters". Finally, there also exists "observation filter" coming from the subjective and objective barriers that make it difficult to correctly identify essential components of reality.

An effective information manipulator can therefore use these filters to influence a group's thinking or perception of reality. This is particularly problematic due to that fact that disinformation may be based on:

- Real information provided in a manipulated context, for instance an authentic photo with a fabricated description.
- False information provided in the right context known to contemporary media, for example dramatized photos that "sell" the atrocities of real war.
- Information, images, films, and recordings that are entirely fabricated, otherwise known as *deep-fakes*.

These types of disinformation can be spread by media masquerading as credible news agencies. In some cases, they are also spread by credible new agencies that are unaware of the content's manipulation (Nemr & Gangwar, 2019). Understanding the context in which disinformation affects social groups requires understanding the failure mechanisms within the human way of thinking. Thinking errors result from the natural human tendency to reduce effort and maximize benefits (Lewis, 2016). In a sense, the human mind is not naturally inclined to conduct in-depth analysis detached from its experience, instinct, or developed techniques for explaining the world. All this is intensified, for example, under time pressure, an excess or lack of data, the recipient's laziness or technical problems accessing verified knowledge.

These recognized and relatively permanent forms of human analytical handicaps, which may be demonstrated both when acting in an individual capacity as well as in the service of strategic analysis for the state, are called *heuristics*, or simplified inference rules. Researchers from various fields have described over 200 such tendencies that represent an immediate risk of creating a cognitive bias or increase in susceptibility to disinformation (Tversky & Kahneman, 1973; Kahneman, 2011; Gilovich et al., 2012).

American authors, in an extensive study of the psychological determinants of disinformation, identified four of the primary mechanisms that explain heuristic approaches:

1. The limited output ability of the human mind to process an increasing amount of information. This is conducive to the simplification of cognitive procedures and intellectual standards, and thus increases susceptibility to errors and absorption of disinformation perceived as true information.
2. Cognitive dissonance results in striving to ignore or minimize information contrary to established knowledge. Disinformation takes advantage of this weakness by profiling the message in such a way that it satisfies the recipient's sense of comfort and does not force them to verify data and change their view.
3. Group pressure and group think. This involves favoring the acceptance of messages affirmed by a given community and increasing the credibility of messages coming from within the group. It is conducive to the emergence of "information bubbles" recognized and used for disinformation purposes.
4. Emotions and the impact of information on the stimulation of actions by recipients. Messages that improve mood and stimulate action (through extremely divergent feelings) often have a stronger impact, thus incentivizing the use of disinformation (Wolters et al., 2021, pp. 24–51; McBride et al., 2021).

Cognitive processes can also be disturbed in other ways. This can be due to ignorance or rejection of knowledge by the decision-maker or the result of intentional and systematic efforts to mislead them. Examples of both types of behavior can be seen in the activity of COVID-19 vaccine opponents. Their rhetoric called for the individual freedom to choose and a respect for human rights. However, at the individual level, resistance to vaccination was largely based on ignorance that multiplied the fear of the unknown. It also increased distrust towards authorities and doctors, who used scientific knowledge contained in a simple empirical message to communicate that vaccination saves lives. At the level of the organized anti-vaccine movement, informational influences aimed to undermine trust in authorities, medical knowledge, and state institutions. They did this by spreading purposefully prepared myths (e.g., "vaccines are made from human fetuses" or "thought-control chips are implanted with vaccines"), manipulated statistical or medical data, and incorrectly generalized truths (e.g., framing cases of allergic reactions to vaccines as evidence of the harm they cause to human organisms).

People often over-estimate their own assessment competencies, thus creating situations that can be exploited by disinformers. There are several simple techniques that exploit human cognitive deficits, described by various conceptual categories classified as:

- Fabrication: deliberately falsifying information to change the recipient's perception.
- Provocation: deliberately falsifying information to trigger specific behavior.
- Silting: overloading a recipient with information to make them unable to properly assess, select, and process it in time.
- Inversion of proportions: providing information in an inverted order of importance, or minimizing the most important information by presenting irrelevant segments as important.
- Dimming: providing information or interpreting ambiguous values in a way that is intended to confuse a recipient or suggest a desired interpretation.

The outlined techniques exploit identified weaknesses of cognitive processes. In complex situations, humans often have difficulties in distinguishing truth from falsehood and consolidating appropriate procedures that allow for the minimization of errors. Often, they succumb to emotions and attempt to reassure the truth of earlier judgments. The constant and daily overload of information that characterizes the modern world favors indifference and almost forces the use of simplifications and cognitive shortcuts. Here, opinions supported by even untrue facts or interpretations that seem objective will have greater persuasive power than pure and dry truth.

Information producers make a deliberate choice about how to respond to two types of situations:

- When there appears a demand for verified and true information. This happens, for example, in the context of school and social education conducted by reliable broadcasters of information, in the activities of honest media, or in the activities of the state, particularly in situations that require high responsibility for decisions.
- Situations that create temptation for adulteration and deliberate disinformation. This includes marketing, election campaigns, political and armed conflicts, and international rivalry.

At the level of the information ecosystem, there is a natural, almost symbiotic coexistence of information truth and falsehood. They are sometimes difficult to distinguish without increased attention and analytical effort on the part of the recipient. This coexistence is favored by modern information technologies and media that accelerate the production and circulation of information. Their sheer volume reduces the chances of verification and, to some extent, prevents people from practicing mental hygiene.

This statement is illustrated by the results of the Eurostat survey from December 2021, which shows that, on average, only 23% of adult citizens of European Union countries regularly verify data they access on websites. The leader in this field was the Netherlands (45% of people), while Lithuania

(11% of people) exhibited the lowest levels of data verification. For Polish citizens, this figure stood at 16%. At the EU level, this survey revealed that approximately 75% of EU citizens do not verify the sources or content of the information they receive (*How many*, 2021). The study also indicates a generational dimension to the problem of disinformation; one indicated that internet users over that age of 65 are seven times more likely than younger users to spread false information on the internet.

Regardless of the form in which it is contained, information is one of the most important goods determining the survival of individuals and social groups, as well as their quality of life. Correct and true information is indispensable for making decisions aimed at satisfying individual and collective needs. It enables rational behavior, an efficient use of resources, and a maximization of benefits. True information in its pure form is one that is not distorted by deliberate or unknowingly duplicated external influences.

5 Infodemia

The key characteristics of the contemporary international information environment is undoubtedly the ever-increasing amount of uncontrolled information produced and transmitted using modern internet-based media, with each of the internet's more than 4.5 billion users able to act as both a producer and a consumer of information. The technological acceleration of humanity, expected to result from the further development of quantum computers and artificial intelligence, will only increase temptation in this regard. It will also provide new opportunities for entities operating in the domain of disinformation to achieve political and business benefits as well as harm individuals, nations, and the international community.

This phenomenon has been aptly defined by experts as an "infodemia", a "global information pandemic", "information smog", "global information pollution", an "information disorder", "infoxication", and "information overload" (Gross, 1964, vol. 2, p. 856). These terms signal not only an unimaginable amount of information but also describe how the information system has become "clogged", resulting in the loss of societies' abilities to process the constantly growing mass. As a result, the quality of decisions made by those societies are deteriorating.

This *infodemia* is distinguished by several characteristics.

First, there is an unknown amount of information currently in circulation. The quantity of this information can no longer be expressed by arithmetic measures conceivable for the average person but only by estimated IT measures such as decimal yottabytes.

Second, digital communication tools and media platforms are proliferating. This guarantees a continuous increase in the quantity of available information, which continues to produce additional "information noise".

Third, there is a rising risk of social actions based on apparent, falsified, or manipulated knowledge, either intentionally or unknowingly.

Fourth, there is a constant responsibility for recipients of information to ensure their own information security and resistance to disinformation. Among an uncountable mass of data, contemporary societies must develop the ability to correctly select the key information needed to make rational individual and collective decisions. At the individual level, this skill is rooted in education. At the community level, it relies on an appropriate system of solutions based on law, international cooperation, regulatory mechanisms, and technological solutions.

The aforementioned infodemia affects the "purity", or natural truthfulness, of messages, which increases difficulty in correctly processing information. Nicholas Carr, the 2011 Pulitzer Prize winner, aptly noted that "when information overload exceeds the mind's ability to process it, the human ability to learn suffers and understanding becomes shallow… it becomes more difficult to distinguish valuable information from garbage, the signal from the noise" (Carr, 2011, p. 125).

In the contemporary world, *contaminated information* has become a common phenomenon, coexisting with reliable knowledge based on verified and true information. Disinformation creates its own "bubbles", gathering recipients that aggressively defend their information space, convinced, for example, that vaccinations are harmful or that the globe is flat. Depending on the topic, such self-replicating "(dis)information bubbles" may have a local or global dimension. Often, they are also of interest to disinformers, who view them as potential target groups. Disinformers work to strengthen distrust towards state institutions and authoritative knowledge within the groups but also use them to undermine the cohesion of wider democratic societies.

Under favorable circumstances, such types of minority fringe groups with parliamentary representation can significantly influence public policy. By expressing their postulates or simply using them to distinguish themselves on the political scene, they may, in certain situations, be able to blackmail fragile coalition parliamentary majorities. An apt illustration of this was the anti-scientific, pseudo-ethical rhetoric of Polish right-wing anti-vaccine parliamentarians during the COVID-19 pandemic (*Nowy show*, 2021). Such attitudes are also represented in the parliaments or congresses of other countries, including in Europe, Russia, and the U.S., as well as the European Parliament, influencing local communities and politics in various ways.

6 The Role of the Media

The media is a key component of the contemporary information ecosystem of states and a potential tool for disinformation (Keane, 1991). It can be classified in various ways (Jakubowicz, 2016), although the advent of social media and online media platforms have integrated what was once divided into traditional and new, into written, audio and video, mobile, and stationary media. Despite debate within the academic community, the media's functions in the modern world are far more important than their formal division (Kozłowska, 2016). The most important of these functions are:
- Pursuing objective and reliable information that reflects reality.
- Selecting and verifying knowledge.
- Transferring knowledge.
- Controlling power and social order.
- Encouraging civic education and support for public debate.
- Acting as a spokesperson for cases of high social importance.
- Providing entertainment (Dobek-Ostrowska, 2004, pp. 135–141).

Fulfilling these roles requires trust that is built over a long time. At opposite extremes, the media can play both the role of a creator of events and an unbiased guide to them, as well as a propagandist who is a party to the issues it covers (McLuhan, 1994). At the same time, the competitive and pluralistic character of the media, which should represent various shades of public opinion in democratic societies, encourages continuous production of content. This, as well as the "24-hour news cycle" element in its operations, is conducive to creating shallow messages that are adjusted to the preferences of consumers. In the face of media that operates 24/7, there is also a need for experts who are ready and willing to comment on any event with little notice, despite a possible lack of knowledge or professional competency in a specific area.

The media plays an important role in ensuring state security in times of war and peace. According to Ociepka (2002, pp. 68–71), to implement the state's foreign policy, the media can be used as:
- A catalyst capable of explaining or complicating the meaning and course of international events.
- A factor influencing the pace of decision-making.

- A tool of propaganda or public diplomacy.
- A form of diplomatic support.
- A decision interpreter and evaluator.
- A factor that determines the order and hierarchy of matters discussed within the public sphere.

In the modern world, social media is increasingly replacing traditional media, and due to its global reach and number of users, it is becoming the main transmitter of information. Because of its decentralized open architecture, the spontaneous nature of information produced on it, and an increasing use of machine duplication methods (bots), social media has become one of the main modern megaphones for information distortion. Of the approximately 3 billion active Facebook users and approximately 350 million Twitter users, almost anyone can, knowingly or not, produce and reproduce false or unverified content. The specifically understood DNA of this type of media allows social media companies to oppose censorship restrictions and flexibly adapt to regulations. These entities are not the cause of disinformation; however, the environment they have created has been instrumental in the dissemination of untested and falsified information (Cosentino, 2020).

CHAPTER 2

Disinformation: Traditions, Lies, Strategy, and Relationship to Politics

Disinformation is a mature, conscious practice that is centered on creating and spreading deliberately produced lies. For millennia, it has been perfected as a communication technique – a tool used to shape interpersonal relations and politics. Within this framework, it is nowadays perceived as a growing threat to the security of individuals, society, states, and international order. Lies, which are the foundation of disinformation, should, however, be considered among the oldest and most natural phenomena that exist in the human environment. They feed human temptation to effectively influence others free from any ethical assessments, and their roots go deep into the past (Philips, 2019). However, the effects of public lies have evolved together with the development of civilization, technological revolutions, and humanity's growing dependence on access to information. So, while it can be said that the phenomena of lies are as old as humankind, modern disinformation has its novelties. Most notable among them is the scale at which the deliberate transformation of information into a tool of social control has occurred, along with the resulting weaponization of information as a strategic resource of the state.

In its most general definition, a *lie* is an intentional communication action in which the liar, be it an individual, state, or media entity, knowingly spreads a falsehood that they themself do not believe (Kucharski, 2014, pp. 93–117). A lie therefore represents a conscious action intended to deceive the recipient. More often than not, the liar has an advantage over the person deceived, who may not have the ability or resources to verify information or has consciously decided not to verify the information. American psychologist Paul Ekman's experiment, conducted on a sample of 12,000 people, showed that half of them were not able to recognize a lie (Ekman, 2014). Only in a fairy-tale was the puppet Pinocchio transparent about his lies – his nose grew whenever he told one. The example of this wooden character is as infantile as it is didactic, but literature is full of cultural patterns of deception. They are not missing in the Bible and play a central role in stories about the beginnings of human kind, and can be found in William Shakespeare's dramas, Fyodor Dostoyevsky's novels, and countless other works.

 | DOI:10.1163/9789004715769_004

The general definition of a lie, however, says nothing about the specific intentions and reasons behind these distortions of information in the communication process. Motives on the liar's side can be innocent, like in the case of a joke made by a diplomat for social amusement, or an undeserved compliment. They can also be noble, intent on sparing a person's feelings. However, history shows that there is no such thing as an innocent lie in international politics. Many years ago, the then Russian ambassador to NATO Dmitry Rogozin sent his Western counterparts a "funny" Christmas card that featured a photo of an Inter-Continental Ballistic Missile (ICBM) with multiple nuclear warheads. The political message was both clear and contradictory to the occasion and to the expected tone of Christmas and New Year's greetings. Its form also did not arouse any enthusiasm or laughter from the recipients. Similarly, on October 18, 1984, U.S. President Ronald Reagan caused a scandal when he announced in a speech on television that he had decided to destroy the Soviet Union and would launch rockets aimed at the USSR in five minutes. This "joke" resulted in a declaration of the highest degree of readiness by the Soviet air defense.

Lies have therefore functioned in human consciousness from the earliest times, coexisting with the truth, though not necessarily as its absolute and indisputable opposite. From the point of view of (dis)information operations, the analysis of their nature requires understanding the phenomenon and the scale of the threat they pose.

1 The Philosophers of Truth

Since ancient times, the concept of lying has also been subject to reflection by philosophers who have considered the coexistence of truth and falsehood through various ontological and epistemological lenses (Kirkham, 1992; Kuenne, 2003). The achievements of these thinkers cannot be summarized with simple conclusions that evaluate the very existence of a lie on the axis of good-evil. For if everything is a *being*, and this originates in thought, then all thought as *being* is equal; hence there is no need to inquire about its truthfulness and the reasons for which it was created. This reflection, though philosophically sophisticated, is only a reflection of reality. For in this, the truth objectively coexists (for some, it struggles) with its deformed representation, which constitutes independent social beings and objects of inquiry. Ancient seekers of truth and perfection went even further. They questioned the very existence of a lie in the face of their inability to undeniably prove what is true and in view of the understanding that a lie does not always mean evil itself. This is the Faustian dilemma of power desiring only eternal evil but doing eternal good. It was demonstrated masterfully by Russian writer Mikhail Bulgakov in *The Master and Margarita.*

It can be argued that the ancient sources of the history of thought (including Plato's works) do not provide an unequivocal explanation of the nature of a lie or a clear condemnation of its active use. They do, however, recognize the universal nature of the phenomenon and various reasons why people and nations mislead each other. It is therefore unsurprising that these types of early philosophical writings spurred later praise of utilitarian attitudes proclaiming consent to falsehood as: an act of reconciliation with human nature; the art of effective governance (like in the case of war when used to raise the morale of a nation or deceive an enemy); or the byproduct of ethical motivations to do good, such as in the case of saving a life or protecting mental well-being.

This direction of reflection owes much to Aristotle, who made one of the first categorizations of disinformation based on lies, which he defined as the act of deliberately misleading someone. Considering it evil, he differentiated motives behind lying, understandingly belittling the guilt of reckless people and seekers of fame and sharply condemning those who did so out of a desire to get rich. Plato's disciple can also be considered an important source of the modern classification of disinformation tools. He noticed a phenomenon recognized as the category of people who cast themselves (sometimes unconsciously) in the roles of "useful idiots", spreading falsehoods due to naivety, a lack of knowledge, or insufficient reflections about the surrounding world (Aristotle, 2002). Such useful idiots are an integral part of any disinformation operation.

This mental impotence was overcome thanks to the biblical tradition and the achievements of Christian writers and philosophers. Much of the "analysis" of this phenomenon was performed by Saint Augustine of Hippo and Saint Thomas Aquinas. While the latter rigorously rejected lying even as an excuse for acts based on noble motives, including saving a life, the bishop of Hippo authored a kind of early study of the morphology of this problem. As someone unequivocally on the side of the truth, he distinguished between disseminating false messages (*lying in speech*) and lying (*lying in deed*) (Goliński, 1936). In his classifications, he designated exceptions for situations in which certain statements, like jokes, even if overtly untrue, did not indicate fraudulent intention. There was much to the understanding of human weakness, although the aforementioned Augustinian reflection runs against the tide of contemporary analysis, incorporating jokes, mockery, and related forms of communication into the (intentional or otherwise) domain of disinformation. This is discussed in further detail later in the book.

Saint Augustine's rejection of lies based on a moralistic interpretation of the Bible was an important point of reference for the later teachings of Christian authors and the Catholic Church. The bishop of Hippo also had his critics, however, for example his contemporary John Cassian, who promoted the idea of the "useful lie" and justified it through historical assessments of events described in the Bible. Saint Augustine's theorizations can also be considered an unconscious precursor to modern definitions of disinformation. He saw two distinct and integrally connected features in a lie: (1) material falsity of the content of the statement, of which the liar is aware; and (2) an intention to mislead the other party (Chudy, 2003, pp. 152–154). This formulation led him to create the first hierarchical typology of this phenomenon, contained in the treatise *de Mendacio* (On Lying), written around the year 395 and further developed in the quarter-century later work *Contra mendacium* (Against Lying). In both volumes, he describes a wide range of circumstances in which a lie may appear but designates the different levels of severity and evil it may carry.

According to Saint Augustine, the most severe type of lie is characterized by a falsification of God's truth. In descending severity, lies are also designated based on the following factors: they harm other people, they are perpetrated for the pleasure of oneself or others; or they do not harm others and are perpetrated to save someone else's property or life or protect them from rape. This view sets a moral foundation for condemning lying as an intentional act but also provides for dual assessments of the gravity of this problem in specific and ambiguous life situations.

The writings of the Church Fathers and their epigones and polemists moving in the opposing space of good and evil, relativism, and cognitive rigorism undoubtedly shaped the Western worldview of lies and their effects on human life. However, teachings based on biblical records and the eighth commandment (*Thou shall not bear false witness against your neighbor*), understood as a condemnation of lying in all its forms, "have suffered a complete didactic failure" (Kucharski, 2014, p. 13).

2 Grotius, Machiavelli, Diplomacy, and the Right to Truth

Modern reflections on lies and disinformation owe a great deal to the Dutch philosopher and lawyer Hugo Grotius. He approached the topic in an Augustinian spirit, classifying truth as a good violated by those who intentionally mislead others; however, he did so in a more secular, earthly order. In his understanding, lying is a form of violence that imposes on others an understanding that differs from their own. In his trilogy of books entitled *On the Law of War and Peace*, he acknowledges the ambiguity of contexts in which a lie can function, but he also considers them an attack on the human right to truth. Admittedly, the right to truth is not, in his opinion, an unconditional right, as it does not apply in:

– Everyday situations towards sick people incapable of the truth.
– The case of joking and accepting untruths in social situations.
– The context of sovereignty – authorities have the right to their own truth, to shape their sovereign communication space and to dictate the rules that govern it, including in the realm of relations with other countries.
– Situations where the protection of good justifies a counterfeit equal to it, otherwise known as a *useful lie*.

The importance of Grotius' teachings is not diminished by the changes in standards of contemporary democratic societies regarding the last two situations described by him. The idea of truth as a human right has been developed in contemporary international relations studies and acts of "soft law" concerning the impact of disinformation on various spheres of rights and freedoms. It includes human rights; individual freedoms; the right to truth and freedom of expression; privacy; and cultural, economic, and political rights, including participation in public life. These issues are reflected in numerous reports and resolutions by the Council of Europe, the United Nations Human Rights Council, the United Nations High Commissioner for Human Rights, and the European Commission, among others.

The reflections of Grotius also influenced the development of political thought and the practices of states operating under the post-Westphalian order that sought to develop legal protection for their interest in mutual relations. Diplomacy became an important instrument for this purpose, undergoing dynamic institutional and organizational development in recent decades, including an improvement in the skills of diplomats. This expansion of skills was based on effective communication, for which the point of reference was the dichotomy of "truth expected" by partners and the "possible truth", meaning various forms of its deliberate distortion used to pursue one's own interests (i.e., propaganda and disinformation).

Niccolò Machiavelli, who lived a century earlier, took the use of truth and its nuances to advise rulers and diplomats in their own interest even further than Grotius. In the interest of politics and achieving its goals, such as the survival of the state and its center of power, he deemed lies beyond moral judgment. In book XVIII of his treatise on governance, he provides a solid foundation for treating lies as a tool that can be used depending on needs and circumstances. He even goes as far as to say they are necessary in environments where political power is rife with falsehoods (Machiavelli, 2014). For Machiavelli, the key to success was not the power of truth or Christian ethical evaluations of an argument but rather the skill and efficiency in using messages elevated to the rank of a strategy (strategic art).

Efforts by Machiavelli, Grotius, and Saint Augustine in assessing the social contexts of lies was further developed by Ambroise Guillois, a Catholic priest and 19th-century commentator on theological texts. Guillois broadly commented on the role of deception in diplomatic communication, building on the foundations laid by Augustine's condemnation of lies. He writes:

> There are circumstances, however, in which ambiguous expressions may be used; for example, someone out of curiosity asks you a question that you cannot answer clearly and firmly without risking indecency or annoyance; on the other hand, your silence would be sufficient to reveal your thoughts: then you can answer with an ambiguous answer to the obsessiveness of the one who asks you ... they are accepted by custom, and their meaning is known. The servant says that you are not at home, although it is real: is he lying? No, because it doesn't cheat whoever it talks to. So, it is not known that these words mean: that you do not want to see anyone, that you do not receive guests? The significance of such an answer is known; no one will be mistaken in it; then you have no lie here. You ask a friend to borrow money; because he knows that you like your expenses too much, and he replies to you: I don't, and yet he has money. Are they lying? No, because the words he says cannot deceive you and only mean, according to the customary practice: I don't have any money that I would like or could lend you. (*Disfavoured truths*, 2018)

It would be a significant oversimplification, however, to regard the views of Machiavelli or the peculiar explanations of Guillois as the sole key to comprehending contemporary diplomatic communication. On the one hand, the history of diplomatic communication is replete with instrumentalized falsehoods and a flexible attitude towards truth, deception, and disinformation. Sir Henry Wotton, the English ambassador to Venice at the turn of the 17th century, characterized the heads of diplomatic missions as "good people, sent abroad, lying in the interests of the country" (Freeman, 2009, p. 9).

Notwithstanding popular opinions and *bon mots* suggesting otherwise, lying is not the foundation of a diplomat's work in representing their country. Communication based on truth is the bedrock of international relations, even in the presence of ubiquitous falsehoods. Occasionally, in times of conflict or permanent interstate hostility, the use of lies or systemic disinformation may seem necessary, but it comes with measurable costs and does not bode well for long-term success. The price of lies in diplomacy includes the erosion of a state's credibility and the credibility of those acting on its behalf, perpetuating conflicts and distrust, and the inability to solve problems that require a minimum amount of honesty. Even if honesty, openness, and truthfulness may not always be a diplomat's primary virtues, lying outright is neither their preferred tool nor a valued aspect of their role.

The fact remains, however, that distrust is one of the most important principles in international relations. While this feature has been exposed by researchers in their attempts to develop a realistic paradigm for the study of international relations, it only reveals the stakes for which the international game is being played: survival, security, and the power of the state. Considering this scenario the norm, it becomes notable when the diplomacy of states becomes systematically dependent on the use of lies and everyday disinformation to hide their intentions or justify their decisions. Such is the case for Russia, for instance, which used these types of tactics to justify its aggression in Ukraine and the annexation of Crimea in 2014 on "defensive grounds". Furthermore, a significant part of Russia's foreign policy employs mythical stories that combine truth and falsehood, framing itself as a country betrayed by its Western partners that must defend itself against a repeat of such treatment (Kupiecki & Menkiszak, 2020; Legucka & Kupiecki, 2022). The mythology of "national humiliation" also appears in the justifications of Chinese foreign policy, which presents China as an ancient civilization mistreated by foreigners for centuries. This narrative asserts that, as a modern-day superpower, China has the right to protect itself against the return of such a situation and should take its rightful place in the world.

Yet this is a typical situation for revisionist powers with aspirations for which disinformation becomes their own version of the truth in communications with foreign states – an operational tool into which the entire state apparatus is harnessed. It is a political strategy capable of effectively influencing the behavior of others while being less costly than war and not bearing the direct risk of confrontation. Such an instrumentalization of disinformation is reminiscent of the old Talleyrand maxim that "in politics, what is believed is more important than what is truth". However, it requires a well-organized state apparatus, a high level of social control, legal regulations, and suitable material resources. Such means are often not available to smaller (and democratic) countries. If disinformation occurs in their activities, it is an exception rather than the rule.

3 Sun-Tzu, Clausewitz, and the Strategic Nature of Disinformation

In the service of state interests, lies and disinformation are not only the concern of ancient philosophers – they are also deeply ingrained in the literature of strategic studies. Unlike philosophers who seek good, truth, and life harmony, military strategists and historians view the manipulation of truth as a legitimate means for governments to pursue their interests. Manipulating the truth is freed from moral considerations and evaluated solely in terms of its potential effectiveness, benefits, and costs, as well as the circumstances that make it possible or difficult to apply. This is not surprising given that the point of reference for these strategists' deliberations is war, which is decisive for the existence of nations. As the popular and apt maxim attributed to the Athenian tragedian Aeschylus proclaims, "in war, the first victim is truth".

Sun Tzu, a Chinese general and a master teacher of strategic thought, emphasized the importance of misleading the opponent in his seminal work on the art of war (Sun Tzu, 2005). Other classics of strategic thought, from the Carthaginians, Macedonians, Greeks, and Romans to Nicolò Machiavelli, nuclear-era strategists, and contemporary theorists of "strategic disinformation", have also drawn attention to this way of using information. The art of effectively disinforming the enemy; blurring their situational awareness; and influencing their decisions, morale, and planning ability are highly valued in the arsenal of strategy. It is also a tactical key to achieving victory at the lowest possible cost in terms of effort and resources, or even without a fight. Such indirect influence on the opponent can cause uncertainty, force them to act defensively, or take away their fighting spirit, all of which weaken them. However, it also hides a deeper strategic reflection, making indirect action a desirable policy tool. It can also win over allies or help manage alliances in conditions of disagreement about the war plan. Indirect actions are a form of combat, which in an optimal situation does not require classic warfare.

Even earlier than the teachings of the Chinese general Sun Tzu, the most well-known example of a mythical message about the application of such an approach is the Trojan Horse. The defenders bringing the horse into the city was preceded by a disinformation operation about the withdrawal of the Greeks, who were besieging Troy. This prompted the inhabitants to abandon their vigilance, which until then had guaranteed their success. The Trojan Horse is arguably the most popular example of an effective hybrid tactic – a disinformation-backed war deception that uses diversion and asymmetric measures. It has also become a cultural synonym for dishonest gifts, and it is no coincidence that it is the name of a malicious digital tool used to steal data. A similar though less known method of conquering the besieged Yaffa was used even earlier, around 1500 BC, by the Egyptian pharaoh Tuthmosis III. He misled the defenders with false information about the withdrawal of troops and sent war contributions in huge wicker baskets where Egyptian saboteurs were hidden in the place of gold. Like the Greeks in Troy, the Egyptians successfully attacked the city at night from within.

A contemporary adaptation of Sun Tzu's strategic thought is the Chinese concept of unlimited war, which relies on disinformation and follows Mao Zedong's theory of revolutionary war (Commin & Filliol, 2015). According to the authors of the monograph that describes its framework, China should use, without any legal or moral restraints, all methods of weakening, exhausting, and defeating stronger opponents. This assumption is primarily related to the power of the United States and the international order as it was shaped after the end of World War II. The arsenal of unlimited war measures includes classic combined military and paramilitary operations; psychological deception; informational and cybernetic operations; and financial, economic, criminal, and terrorist instruments (Liang & Xiangsui, 2002). The aforementioned monograph, which first appeared in its Chinese edition in the late 1990s, may have been a harbinger of an idea in the form of guidelines for the Chinese armed forces known as the "concept of three wars", or three types of warfare. This concept assumes a combination of informational (shaping public opinion at home and abroad), psychological (influencing decision-making processes), and legal (building legal arguments supporting the implementation of the state's interests) means of strengthening China's international position (Mattis, 2018; Spalding, 2022).

For Carl von Clausewitz, a classic of modern strategic thought, disinformation represents more than just Sun Tzu's ideation of a tool of war – according to von Clausewtiz, it is also a tool of policy. For him, in a conflict, information and disinformation coexist on an equal footing. In his understanding, warfare is always carried out "as if in darkness, […] which, like a fog or moonlight, gives things an exaggerated size and a bizarre appearance" (von Clausewitz, 1984). This "fog of war" is an objective reality that increases uncertainty for commanders and politicians alike. While opponents can magnify this uncertainty, disinformation as a combat tool must be planned precisely, with a clear understanding of its benefits and costs. The key to success lies in identifying the opponent's "center of gravity", which, when hit, will lead to their defeat. Clausewitz's concept of "center of gravity" is the essence of effective action, both in times of war and peace, and requires a constant and careful analysis of changing circumstances.

While modern Chinese strategy heavily draws on the thoughts of Sun Tzu, Russian strategic thinkers and their followers in the 21st century are students of both the ancient Chinese general and von Clausewitz. From both, they draw an understanding of the integrity of space and domains of conflict, which does not distinguish between military and non-military domains; traditional operational domains such as land, air, and maritime versus extended technological, cyber, and information battlefields; or the separation of times of peace and war (Darczewska & Żochowski, 2017). They also see new asymmetric possibilities of effective informational influence on Westerners to weaken the cohesion of democratic states and limit their resilience. Without social support or in conditions of divided public opinion, it can be difficult to govern effectively; in such cases, making difficult and bold decisions may turn out to be impossible or even deadly for democratically elected authorities. Russia's reference point for such actions is the concept of *political warfare*, which was defined by George Kennan as the:

> logical application of Clausewitz's doctrine in peacetime [...] application of all means at the disposal of the state below the threshold of war to achieve national goals [...] in an open and secret manner, from such explicit measures as alliances, economic and "white" measures propaganda, to covert operations to support "friends" abroad, "black" psychological actions, and even incitement to resistance in hostile countries. (Kennan, 1948)

Disinformation is important and facilitates linking various measures into complex scenarios of interaction, but it is only one dimension of contemporary activities referred to as political warfare, or the *fourth generation of warfare* in Anglo-Saxon strategic studies (Robinson et al., 2018, pp. 2–6). It features a particular mutation of peacetime in the form of *hybrid warfare* or a *gray zone conflict*. This concept includes: (1) coordinated and synchronized actions, using political, economic, military, civil, and information measures against the weaknesses of democratic states and their institutions; (2) difficulty in detecting or attributing the actions to specific actors due to their operation in a space between peace and war (i.e., the gray zone); and (3) the perpetrator's aim to influence various decision-making processes in pursuit of their own interests (*The Landscape*, 2021).

These concepts often lack precise and universal definitions, regardless of whether they are viewed through Kennan's categories of mobilizing all components of the national potential to act below the threshold of war or in Clausewitzian terms, which see war as a political tool. The author of the term "hybrid war" admitted to taking the idea of the waning boundaries of conflict and the loss of the state's monopoly on violence from the concept of fourth-generation warfare. Furthermore, the concept of multi-domain warfare was borrowed from Chinese proponents of unrestricted warfare, net-centric warfare was taken from American strategists, and the synergy benefits of combining conventional and unconventional capabilities at a lower and more integrated level comes from the proponents of *compound warfare*. The complex and dispersed nature of the operational environment and the opportunistic nature of future adversaries were derived from Australian experts (Hoffman, 2007, p. 30).

The lack of precise definitions is because the focus remains not so much on the classical understanding of military operations but instead on treating these actions as interconnected and non-kinetic activities that construct various scenarios of conflict escalation below the threshold of open war. These scenarios aim to achieve goals through information manipulation and multi-level non-military activities while avoiding open clashes of troops. The end phase of such scenarios may lead to a full-blown conflict, but the key is to achieve objectives through non-military means. In non-military activities, it is difficult to identify the perpetrator or assign responsibility in accordance with international law, which would imply, for example, the right to retaliate (Treverton et al., 2018, p. 10).

In times of peace, controlling information – whether it is true, partially distorted, or false – can help create an image of power that is not worth challenging. This can help convince opponents that aggression is not a profitable course of action. The impact of this image depends mainly on the strength of persuasion, and, to a lesser extent, on the limited possibilities of confronting it with reality. This is particularly true when it comes to communicating data, such as the possession of military capabilities and their potential use in specific situations. The effectiveness of a deterrence policy, for instance, relies on this kind of behavior (Lorenz, 2021, pp. 15–92). Extensive mechanisms of strategic communication supported by intelligence organizations and public diplomacy are employed to achieve this, drawing on knowledge from social sciences, humanities, and modern mass communication technologies.

4 The Colors of Propaganda

Diplomacy, strategy, and the art of war are linked by a distinctive, active way of using information with the intent of communicating one's intentions, hiding them, or influencing the perception of partners and opponents. This is hidden under the concept of propaganda. In a linguistically neutral context, propaganda means communicating and promoting specific messages. The concept itself indicates only the operational mechanism of action, saying little about the content being promoted or the specific intentions related to it (Cunningham, 2002).

In this sense, propaganda can be a tool for promoting negative agendas, such as encouraging genocide, creating a cult of personality of leaders in totalitarian states (Kupiecki, 1993), or justifying aggression and territorial conquests. It can also contribute to positive causes, however, such as vaccinating children, promoting occupational health and safety, and advancing cancer prevention. In international politics, propaganda can be used to share good practices in governing and organizing elections.

However, the political practices of states, international organizations, and businesses have given the concept of propaganda a negative connotation. It is most often associated with information manipulation, *spin*, dishonesty, brainwashing, and psychological operations. The classical definition of propaganda defines it as "the art of influencing, manipulating, controlling, promoting, changing, encouraging, or supporting opinions, attitudes, actions, or behavior" (Martin, 1958, p. 10). According to the authors of the most widely read and repeatedly re-published textbook on propaganda analysis, propaganda assumes the purposefulness of an action which aims to maintain or change the balance of power in relation to the recipient of the propaganda. Jowett and O'Donnell (2015) define propaganda as "a deliberate, systematic attempt to shape perception, manipulate cognitive processes, and direct behavior to implement the propagandist's intention" (pp. 4–9). Simultaneously, they distinguish between propaganda as an act of manipulating the recipient of a message and persuasion based on a relationship and mutual perception of benefits for the sender and recipient.

In this realm of meaning, relationships, and communication, Jowett and O'Donnell (2015) distinguish between two types of propaganda: agitation and integration. Modern states employ both forms of informational interaction, using propaganda and disinformation to achieve their goals. Agitation involves using words and images to move the recipient and persuade them to act in a specific way, changing their behavior or support for certain actions. This type of influence is apparent in the activities of anti-vaccination movements, for instance, as well as in Russian information operations that use the justification of necessary defense to rationalize their foreign policy (Kupiecki, 2019). Integration content, on the other hand, creates messages with the aim of strengthening social acceptance and affirming the cause around which the propaganda effort is built. Examples of integration propaganda include China's attempts to shift responsibility for concealing information about the spread of COVID-19 (Dubow et al., 2021), Donald Trump's domestic propaganda built on the slogan of "making America great again", and Russian efforts to consolidate public support for the annexation of Crimea.

Jowett and O'Donnell (2015) also recall older classifications of propaganda (Becker, 1949, pp. 221–235; *Foreign Service*, 1951, pp. 955–956) into "white", "gray", and "black" based on the credibility of the source, the nature of the information used, and the persuasive intent of the propagandist (pp. 20–28). Within this framework, *white propaganda* is defined as an act of communication that benefits the sender and shapes their positive image while respecting the recipient's needs. *Gray propaganda* loses this clarity, involving half-truths, falsifications, frequent source concealment, and messages that may harm the recipient's knowledge and attitudes. *Black propaganda* is equivalent to disinformation, in which the truthfulness of the information is irrelevant if it induces the desired effects in the consciousness, attitudes, and decisions of large social groups. The intent to harm the recipient and disregard their needs lies at the heart of black propaganda. In this case, the operator is hidden, and it can be difficult to prove their connections with a sponsoring state.

Among the numerous concepts that have a similar meaning to propaganda and relate to the deliberate and targeted handling of information, it is worth defining the following terms:

- Press and information activity of public affairs. This refers to specialized institutions such as press offices, spokespersons, situational centers, or press attachés. They can be either civilian or military in nature, with the former being related to timely, precise, advanced, or reactive information through the media about the policies and undertakings of a given entity. The latter is associated with external information about the entity's sphere of military activity. The aim of these institutions is to create a favorable information environment that increases public support for the entity's activities. Additionally, these institutions ensure consistency in external communication across all centers of the entity.
- Strategic communication (StratCom). This refers to a field of activity that involves coordinating and applying specific methods of informational influence to the intended recipients at the appropriate time and with the appropriate specificity, to achieve the objectives of a given activity.
- Information operations (InfoOps). This refers to various information activities utilized to attain a desired effect on a recipients' motivation to act, their perception of issues, and their capabilities.

- Psychological operations (PsyOps). This refers to the deliberate and systematic use of information to influence the cognitive processes and behaviors of social groups, with the goal of achieving political and military objectives.

The above definitions primarily reflect the Western perspective on the elements of information warfare. However, their interpretation and practical application differ in countries outside this cultural sphere, which have developed their own approaches to the field.

Information and psychological operations are a specific type of activity that involve a series of undertakings such as support, counteraction, and information defense. They are carried out according to a uniform concept and planned with the aim of gaining and maintaining an information advantage over the enemy during military operations. The purpose of conducting information and psychological operations is to disorganize the function of the enemy's information infrastructure and, consequently, cause its state structures to collapse. The success of the operation depends on constant pressure exerted on the opponent and the maintenance of the psychological initiative.

Scientific analysis of propaganda has led to the classification of approximately 100 techniques of information manipulation that operationalize knowledge about the psychological and cognitive processes of people and the dynamics of group behavior. Conserva (2003) has divided them into the following categories (Cole, 1998; Shabo, 2008; Da San Martino et al., 2019):

1. Applying fallacious logic. This involves drawing false inferences from data that does not back them up.
2. Using eristic ploys and diversion. These include, for instance, *ad hominem* rhetoric, use of content *ad nauseam*, references to emotions, and selective use of quotes or authorities.
3. Appealing to emotions through a "love offensive." This involves an ostentatious showing of interest and care, as well as the use of fear, flattery, hate speech, patriotism, and higher emotions.
4. Using falsehoods and deceptions. This involves the operationalization of lies, half-truths, generalizations, and inverted meanings.
5. Playing a game with human behavioral inclinations and limited inference abilities. This is done by dehumanizing the opponent, demonizing problems, creating group pressure, and causing cognitive dissonance.
6. Using a propagandistic style of speaking or writing. This style includes the use of false accusations, labeling, exaggeration, positive words, minimization, and slogans.
7. Arguing based on the "common sense" argument.

All these propaganda measures are present in contemporary disinformation campaigns carried out in the international arena. They have been proven effective with identified target groups and are able to exploit the perceptive weaknesses of individuals and social groups, which are intensified by an overload of information and its channels. This hinders not only verification but also critical reflection (Kavanagh & Rich, 2018). Some of these measures seem like innocent coloring or "creative" processing of the truth, and in some cases they have even been used as "proof" of the intellectual and social sophistication of their users. A perfect example of this is the media "talking heads" from the world of politics and commentary who efficiently use eristics, once described by philosopher Schopenhauer (1893) and processed by spin theorists. In general, however, these measures lead to distortions of reality, constituting acts of direct disinformation or, by proxy, efforts aimed at lowering public sensitivity to information manipulation.

Contemporary state use of gray and black propaganda includes, for example:

- Traditional state-controlled media.
- Social media, which can be used as an enabler of activities perpetrated by paid, organized disinformation groups (e.g., troll farms) (Hughes & Waismel-Manor, 2021).
- Trolls, i.e., individuals paid to produce content that is biased in favor of their clients' goals – they comment on reality in a biased way, building criticism or support for specific individuals or issues.
- Bots, which are automated information production and duplication processes responsible for the increasing number of false messages intentionally disseminated online. Particularly active on Twitter, bot accounts are often characterized by a lack of identifying or false photo of the owner, a small number of followers and many accounts they follow.
- Specially created organizations that imitate legitimate civil society institutions.
- Secretly sponsored radical and populist parties, opposition movements, and peace campaigners, for example Soviet support for Western anti-war, environmental, and anti-nuclear movements during the Cold War.

As part of disinformation operations, these actors often operate in a coordinated and planned manner, becoming in effect a broadly understood "influence agency." The tools of influence should be considered more broadly, however, extending beyond the strict nomenclature of intelligence services. It is also important to not forget the "useful idiots", an immortal phenomenon of people supporting social and international pathologies for various reasons. Without naming them as such, the "useful idiot" phenomenon was first characterized by Aristotle. The modern understanding of this human tool of manipulation is ascribed to the leader of the Bolshevik revolution, Vladimir

Lenin, who so described the uncritical Western eulogists of the freedom and successes of Soviet communism. Given the lack of evidence as to who coined the term itself, it must be considered a historical construct of scientific debate (Charen, 2003, p. 10).

The perfidy of propaganda and disinformation lies in the accurate recognition of the needs, values, and emotions of the message's recipients. Techniques that reinforce existing beliefs within a specific social group are particularly effective. Over time, repeated interactions based on these techniques lead to an instinctive acceptance of information that confirms preexisting beliefs while rejecting anything that contradicts them, regardless of the source's authority. Furthermore, persuasive propaganda clichés contribute to a shift in cognitive habits towards mental laziness, reducing the tendency to verify information sources or fact-check data.

Such an approach alleviates participants' feelings of isolation, discomfort, and potential shame arising from their ignorance. It also provides a sense of security, identity, strength, and satisfaction, along with a distinct social legitimacy resulting from the existence of a separate communication space with its own truth and significant influence. Such a worldview may lead to political representation for that perspective or its adherents, which can be leveraged by groups seeking to expand their electorate. Maintaining this relationship's consistency may generate impulses to develop disinformation that supports the constitutive elements of an information bubble. This same mechanism can be replicated by foreign states who sponsor such relationships for their own benefit, aiming to enhance divisions within foreign societies and hinder or manipulate their decision-making processes. "Alternative facts," simplistic truths, and black-and-white explanatory constructions that strongly appeal to emotions have significant potential power, particularly in situations where economic or political realities are too complex to comprehend or sudden events occur (Wolters et al., 2021).

Disinformation, driven by lies, plays out in people's minds and is facilitated by modern technology and social engineering techniques. However, the consequences of individual behavior are amplified in large social groups, resulting in visible impacts on states and nations. Within the history of strategic thought, the recognition of information as a weapon in political and military conflicts (i.e., the weaponization of information) does not surprise or provoke objections. The essence of the deliberate use of information in times of peace, crisis, and war is to influence the perception of reality and the resulting actions of an unwitting target against whom operations are carried out. Information used as a weapon for disinformation purposes therefore works against its intended target and benefits the disinformer.

The contemporary connotations of this concept, which are cognizant of disinformation interactions occurring in an environment saturated with digital technologies and internet-enabled services, are sometimes described as "hacking cognitive processes" (Mann, 2008). However, disinformation primarily exploits human weaknesses in terms of how people learn about the world, their deficiencies in critical thinking, and their individual and collective attitudes toward problems and values, which can be culturally, psychologically, or socially conditioned (Erbschloe, 2019). In this context, technological measures only act as intermediaries between the sender and the recipient, serving the purpose of transmitting information and shaping or distorting the information space.

While the phenomenon of weaponizing information is as old as humanity itself, the term itself is relatively new. It belongs to the same generation of concepts used to describe the tools of contemporary international conflicts, such as *lawfare* or *hybrid warfare*. While it's challenging to pinpoint the author of this concept, all evidence suggests that it emerged from Western analyses of Russian information operations during the illegal annexation of Crimea in 2012–2014 and its subsequent actions against Ukraine (Pomeranstsev & Weiss, 2014). In this context, the comment attributed to Russian defense minister Sergei Shoygu, who said that "words also shoot", are particularly relevant. Today, however, the use of information as a weapon is associated with the activities of other undemocratic regimes, including China and Iran, as well as non-state actors. This classification of information is determined by its intended use and the way it is used to the detriment of the recipient.

CHAPTER 3

Disinformation: Conceptual Framework and Origins

Defining disinformation poses two main challenges. The first challenge is that researchers and practitioners use multiple synonymous or even identical terms, the most common of which are:

- "Disruptive communication" or "disturbed communication", both of which describe information that contradicts its goal of conveying objective information (Bennett & Livingston, 2018).
- "Propaganda", whether gray or black, is often used as a synonym for disinformation.
- "Clickbait" is a form of information manipulation. In a narrower sense, it represents a keyword or headline with intriguing content designed to persuade the recipient to "click" on a link.
- Several other terms like "biased", "untrue", "partially true", "social engineering", and "persuasion" indicate the nature, purpose, or use of distorted information.

The relationship between these terms and the studied topic is indisputable, as is the rich history that accompanies each of them. They refer to the systematic "propagation of messages with characteristics of information prepared in a biased manner in order to elicit specific behaviors or attitudes" (Mazarr et al., 2019). This method of influencing another person, groups of people, or the state in relation to another state is a constant component of the reality in which various entities compete for specific goods such as international position, security, control over resources, and military victory. In this context, a lie can function:

- As a separate policy tool, referring to a specific matter, person, or group of people.
- As a form of interrelated narrative sequences, namely the course, evaluation, or interpretation of more complex phenomena, sequences of events, or social processes.
- As a component of complex operations, in which information plays an important role in paving the way for the use of other policy tools.

The common elements of these terms are:

- Their use of social tools that allow for manipulative influences persuading decision-makers to act in a specific manner.

 | DOI:10.1163/9789004715769_005

- The voluntary behavior they induce.
- Their deliberate attempts to harm an object of disinformation.
- Their awareness that the effects of their actions may not be in the best interest of the decision-maker but rather benefit the interests of the source of manipulation (Afeltowicz & Pietrowicz, 2013).

The decision-maker can be an individual taking action, for instance a person voting in democratic elections. It can also be a state body or a social group exerting pressure on constitutionally empowered state authorities. Depending on its purpose, the importance of the goals it supports, and the complexity of the mechanisms serving it, disinformation can therefore be divided into three categories:

1. Tactical. This is immediate and used for the simplest situations, such as inducing a specific one-time behavior change.
2. Operational. This form is more complex due to the nature of the goals it serves. It functions over longer or repetitive time sequences, relating to multi-thread issues that do not constitute an individual decision-making situation. It may concern, for example, undermining trust in democratically elected authorities or scientifically proven knowledge.
3. Strategic. This category of disinformation serves the achievement of long-term goals concerning the most vital issues, which are precisely defined by the dispatcher of disinformation activities. An example would be causing a permanent change in recipients' ways of thinking or evaluating phenomena.

Through the use of disinformation, states can offensively influence the communication and decision-making processes of other international actors. This can result in desired changes, legitimization of their own goals, consolidation of their own community, and reduced costs in comparison to other methods. Disinformation can also be used to defend international decisions and behaviors, hide or falsify actual intentions, and craft content to build a friendly environment. States can also employ a hybrid strategy that integrates many internal and external as well as offensive and defensive goals at the same time.

1 Definitions of Disinformation

Terms such as "alternative facts" and *"fake news"* were recognized in 2017 by the Collins Dictionary as words of the year (Fake news, 2020) and they have gained in popularity in recent years. The use of such language sometimes leads to linguistic absurdity, however, as reflected by the aforementioned "alternative facts". This concept is an obvious oxymoron, similar to the term "true lie", which also accurately describes the spectrum of misinformation regarding the communicative status of subjective truth. The only value of such frivolous metaphorical language games is in their power to shock recipients and draw public attention to the urgency of the problem they are referring to.

According to McNair (2017), *fake news* means "false information often of a sensational nature published in the media with the intention of misleading the recipient in order to achieve financial, political, or prestigious benefits." Researchers at the Reuters Institute for the Study of Journalism distinguish several categories of information that can be included in this group:

1. Consciously distorted information used to achieve a certain result.
2. Information that is entirely invented to achieve a political or financial goal.
3. Information resulting from the sender's low level of professionalism, including poor techniques, factual errors, messages with misleading headlines, and the use of clickbait titles.
4. Information derived from situations where the term *fake news* is used, for instance by politicians, to discredit the source or the information itself and achieve political goals. An iconic example of such behavior is the media activity of former U.S. President Donald Trump, who accused his opponents and the "unfriendly media" of "spreading fake news", or more simply, "of being fake news". It is estimated that throughout his four-year term, Trump spread more than 30,570 lies (Kessler et al., 2021).
5. Information that looks like reliable journalistic material but is, in fact, advertising material.
6. Invented information used to make an audience laugh or ridicule another person (Newman et al., 2020).

The multiplicity of definitions in scientific literature and state documents, as well as those produced by international organizations, is another issue in the study of disinformation (Jayakumar et al., 2021). Academic definitions often take a broad approach to the investigated phenomenon, but individual researchers may differ in their characterization of the problem or its nature. Some definitions emphasize certain details more clearly than others, such as:

- The benefits for the manipulator measured by the effectiveness of its influence on the recipient.
- The purposefulness of the manipulator's actions and the level of message distortion used – the latter does not necessarily have to be based on false or artificially fabricated information.
- The intent to harm.
- The areas of the benefit sought (e.g., political or economic).

The dissimilarity of research approaches, within a relatively coherent understanding of the conceptual framework, often results from the specific interests of individual scientists or the scientific discipline within which the research is conducted. This can sometimes lead to terminological confusions, however, as some researchers use certain terms interchangeably to explain the phenomenon of disinformation, while others insist on the differentiation of meanings and the problems they describe.

An illustrative case is the comparison between propaganda and disinformation. Propaganda has a longer history, with roots dating back to the 17th century, reflected in the programs and institutions of the Catholic Church. Disinformation is a modern incarnation of this phenomenon, incorporating propaganda methods but utilizing more advanced tools and specific goals and strategies. Both phenomena are used in state policy, but propaganda is ubiquitous in the activities of most countries with sufficient resources and organizational facilities, whereas the use of disinformation is closely linked to the strategic goals and nature of a particular state's policy, its operational environment, and its reasons for employing asymmetric offensive methods during peacetime.

Despite their similar origins and use of information tools and techniques, propaganda and disinformation are two different phenomena, particularly when considered from the perspective of their intended harmfulness for the recipient. Propaganda, especially "white" propaganda, which openly presents the information's intention or originator, is a legitimate tool of state policy, including foreign policy, which is often referred to as public diplomacy. Persuasive intention, the desire to present one's case in the best light, or the intention to influence a recipient does not necessarily imply harm. On the other hand, "gray" and especially "black" propaganda can have effects similar to those caused by disinformation. Rajczyk's (2016) monograph on contemporary information wars presents a radical position on this issue, considering disinformation only as a component of propaganda, not an independent phenomenon and subject of research (p. 13). Although this may be true on the semantic level in propaganda research, it does not hold up against a strategically located state disinformation apparatus, placed at the service of the state's policy and developed as a doctrine of integrated action in times of war and peace.

Bennet and Livingston (2018) have developed the concept of *disruptive communication*, which defines disinformation as a policy tool – an "intentional falsification disseminated as information or a simulated series of facts to achieve political goals" (p. 124). Polish researcher Aleksandrowicz (2016), on the other hand, sees disinformation primarily as a method of action, describing it as "a way of transmitting information, whether true or false, to mislead the opponent/competitor and persuade them to behave in line with our expectations for our benefit" (p. 83). In fact, both definitions grant similar social status to true and false information, with the caveat that false information only pretends to be true for political or other intentions. In this sense, true and false information are like gold and tombac – shiny and able to seduce and please buyers.

Vladimir Volkoff, a French researcher of Russian origin, also views disinformation as a weapon and a doctrine of politics and conflict. He believes its significance lies in the space between black propaganda, deception, and information diversion that distorts situational awareness and influences social attitudes and behavior. The mass of falsified information circulating in the attacked state determines its specificity, and effective disinformation only gives shape, direction, and meaning to it after identifying the situation. Volkoff argues that effective disinformation operating in a specific context does not need to be built from scratch but can instead use cognitive gaps or gaps in people's moods and communities that can effectively influence the behavior of large groups (1991b, pp. 5–12).

Joanna Darczewska (2017) from the Center for Eastern Studies in Warsaw proposes a definition that refers to her area of research but has high generalizing value. She treats disinformation as "a collective concept of various methods used in physical and information space, synthesizing political, military, intelligence, business, diplomatic, media, and cyber techniques, which are much more subtle than simple fraud and serve the implementation of long-term political goals". Her definition draws attention to the hidden, indirect, and difficult-to-identify nature of disinformation, understood as long-term indoctrination and destabilization of opponent societies; it has a systemic nature, multi-level interactions, and uses various communication channels, including diplomacy, politics, the economy, the army, and the media. Not every action in the areas indicated by Darczewska must be disinformation, but her definition highlights the fact that in Russian practice and for Russian state-controlled entities, it may become so.

When analyzing individual definitions, a general consensus can be identified regarding the following elements of disinformation:

1. The message is based on false or manipulated information.
2. The intention of the message is to influence the recipient in a way that is objectively inconsistent with their interests and values.
3. The purpose of the action is to elicit a response that is consistent with the disinformer's intent and benefits them.
4. The disinformer does not consider the harm caused to the recipients of their message and the actions taken for their own benefit.

In scientific terms, the definitional approaches mentioned above are reasonable for researching disinformation. However, they may not be useful for state services, international organizations, or bodies dealing with identifying and combating disinformation. Clear and precise definitions are necessary to develop specific and feasible strategies that strengthen social resilience, operationalize planning processes, and schedule activities. This is particularly important when state activities fall within the civil and military spheres, where a precisely formulated planning premise is necessary to manage resources and a shape defense processes in a responsible way.

There are also significant discrepancies in the category of practical operational definitions of disinformation. These discrepancies are partly a result of the problems discussed above but are further exacerbated by the fact that the authors come from various sectors and services responsible for state security. The definitions used in these sectors serve the sectoral conceptualization of the issue, but they do not necessarily result from a comprehensive understanding of the phenomenon from an epistemological point of view. Instead, they are a formula that relates disinformation to the activities of a given service or formation based on their competencies regulated by state law. There is nothing inherently wrong with this approach as long as it aligns with the sectoral operationalization of more general concepts derived from national strategies, especially national security strategies.

A good example of the effort to include disinformation in the broad context of state information security and address this problem is the 2015 *Information Security Doctrine of the Republic of Poland*. However, the definition of disinformation contained in the doctrine is more academic in nature and may not be suitable for the needs of state institutions. The definition reads as follows:

> Propaganda, disinformation – disseminating manipulated or fabricated information (or a combination of both) in order to persuade recipients to certain behaviors beneficial to the disinformer or to divert their attention from actual events. (*Projekt Doktryny*, 2015)

While the suggested definition (which contains references to Russian disinformation operations defined by Alexander Golitsyn, discussed elsewhere in this book) is correct, it appears too general and does not provide a clear picture of expected state actions against the threat of disinformation. Although the document has extended validity and allows for flexible activities of state institutions, it still lacks coherence in addressing the issue. The main objections to this definition can be formulated as follows:

1. It treats disinformation, similarly to propaganda, as a homogeneous phenomenon, neither distinguishing between the various types of actors using this tool for their own purposes nor considering the central role of disinformation in the state strategies of its main actors.
2. Without references to the planning of the disinformer's actions, the definition equates, for example, the phenomena of deliberate disinformation (i.e., the weaponization of information by a foreign state or non-state entity) with the unconscious transmission of unverified information or the transmission of hate speech. This is a definitional issue, addressed through the triad of *disinformation-misinformation-malinformation*.
3. Each element of the above triad is a problem, a potential *sui generis* threat, but they do not necessarily require equal attention from the state. Fighting each of them will have a different priority in the activities of state institutions and services, which will require different tools and allocation of resources.
4. The definition ignores the instrumental role of disinformation as a component or activator in complex hybrid strategies.

Other working definitions of disinformation used in the work of government institutions, such as the Polish Government Center for Security, also demonstrate a similar level of generality and selectivity. The most often used definition, which is a derivative of numerous dictionary approaches, identifies disinformation as a specific type of message based on falsehood, "the purpose of which is to evoke a view, decision, action, or lack thereof in the recipient, in accordance with the assumption of the center that planned the process of misleading the recipient" (Basaj, 2018, pp. 14–17). However, this definition only focuses on a narrow aspect of the tool used, the "message based on falsehood," and does not determine whether every message (regardless of its form and the author's intention) constitutes disinformation.

This definition also raises a question that often appears in strategic studies, which goes beyond technical inquiries. It is about seeing a given phenomenon not only as a tool but also as an organized concept (strategy) applied by a given international entity and closely intertwined with its policy. It refers to a coherent set of goals, an understanding of the security environment, and internally related and repeatable techniques of operation that use rationally selected resources belonging to the entity. From this perspective, disinformation should be considered as:

1. A particular way of formulating and using information with the intention of harming a collective audience and benefiting the sender.
2. A concept of permanent information weaponization and a strategy used both in times of peace and war.

3. A threat to the integrity of democratic societies as well as national and international security.

This approach to disinformation could be informed by examining strategic practices in countries such as Russia and China. However, the *Information Security Doctrine of the Republic of Poland* never came into force; if it were to be used as the basis for a state document today, it would require thorough reflection and supplementation. This matter was not sufficiently addressed in the binding *National Security Strategy of the Republic of Poland 2020*, which does not provide a definition of disinformation. However, it rightly includes disinformation as part of the threats and tasks in the field of state information security, as stated in Section 5.1: "At the strategic level, to build capabilities to protect the information space, including systemic combating of disinformation, understood as interpenetrating layers of space: virtual – the layer of systems, software, and applications, physical – infrastructure and hardware and cognitive" (*Security Strategy*, 2020, p. 21).

In the latter context, the indication of the integral relationship between its material, virtual, and cognitive dimensions should be considered interesting for the future characteristics of the phenomenon.

From an operational point of view, disinformation is much better defined by international security organizations. The North Atlantic Treaty Organisation (NATO) defines disinformation as: "deliberately creating and disseminating false and manipulated information with the intention of fraud or misleading, leading to deepening divisions among allies, which undermines citizens' trust in democratically elected governments" (*NATO's approach,* 2020). This definition clearly indicates the set of protected values as well as the nature of the problem requiring collective action from the allies.

The definition used by the European Union characterizes disinformation broadly as: "false or misleading information created, presented, and disseminated for financial gain or to knowingly mislead public opinion – distorting public debate, undermining citizens' trust in institutions and media, and even destabilizing democratic processes such as elections" (*Questions,* 2018). This definition emphasizes the intentional political and economic motivation of perpetrators of disinformation.

In 2018, the EU created its own Code of Conduct aimed at limiting the scope of disinformation in the field of combating disinformation (EU Code, 2018). Interestingly, in detailed explanations, the EU definition excludes "misleading advertising, misrepresentation of facts, satire, and parody or unequivocally biased media comments". This approach is understandable in terms of the right to freedom of expression or, more generally, freedom of speech. However, it should be noted that these exceptions may become carriers of disinformation and components of information operations. The aforementioned document is not a legal regulation, but an incentive for technology companies to self-regulate and limit the possibility of using social media and the internet in general to spread falsehoods.

An interesting approach to defining information has also emerged in the work of the European Parliament. In one of its studies, the phenomenon was characterized as the sum of four features:

1. The use of falsified, manipulated, or misleading information and unethical methods of persuasion.
2. The targeting of an important topic of public interest.
3. The intentional creation of a sense of threat, hostility, or polarization and the undermining of democratic processes.
4. Dissemination that employs aggressive and automated techniques such as bots and artificial intelligence, trolls, and *micro-targeting* to increase reach (Bayer et al., 2019, p. 9).

The definitions adopted by NATO and the EU institutions regarding the impact of disinformation on state security and international economic relations are key to recognizing the phenomenon itself and understanding its possible implications. Its impact on civil liberties and media freedom is also emphasized by the activities of the Council of Europe and the Organization for Security and Cooperation in Europe (Gerrits, 2018, pp. 16–18).

A foreign state can gain even partial control over the way information is processed by another state's society, including its content. This method of influencing perception can be politically effective, allowing the foreign state to achieve its goals in a less costly manner than through long-term economic and political coercion or armed conflict (Kick, 2001; Melton & Wallace, 2010; Houston et al., 2012; Auerbach & Castronovo, 2013; Levine, 2014; Smith, 2014; Pacepa & Rychlak, 2015; Journalism, 2018; Shu et al., 2020).

Without delving too deeply into definitions, we can aim to create a synthesis that combines scientific reflection with practical usefulness for state institutions. This is the point of reference in this book. Disinformation can therefore be defined as *a doctrine and practice employed by states or non-state actors to deliberately use manipulated or falsified information in order to induce a desired change in a specific audience within a planned area of influence. It is intended to harm the recipients and is used as part of information and propaganda operations, employing techniques of influence and psychological manipulation during times of peace, crisis, and war.*

Naturally, the practice referred to in this definition may influence the definition of the subject of disinformation, which is a general doctrinal sense that is collective. However, in narrower perspectives, such as sectoral, military, or intelligence, it may change to include individuals or smaller target groups. In this form, it primarily relates to the international dimension of the phenomenon, which poses a threat to contemporary democratic states. It should also be noted that there is also a significant amount of disinformation activities directed by authorities toward their own countries.

Disinformation is a tool of political competition in both the domestic and local dimension, and democratic states are not free from them. As argued by Sava Gunitsky (2020):

> They have many incentives for disinformation built into their institutions, perfectly compatible with democratic norms, and factors for building healthy democracies also promote disinformation. The democratic process requires the free flow of information and many competing narratives, and many media and political actors convince their audience in the market of ideas. He tolerates and encourages conflicting opinions, including foreign ones. All these advantages of democracy, unfortunately, also create massive incentives for disinformation.

These observations are confirmed by the report on the global information order published in 2019 by Oxford University, which revealed that between 2017 and 2019, the amount of disinformation transmitted via social media doubled. It also showed that over 70 countries used computer propaganda to manipulate public opinion. Of these, 45 countries were entities operating in democracies and using disinformation in election campaigns to gain support, while 26 were authoritarian countries using disinformation to persecute public opinion and the free press. The preferred tool for these operations was Facebook, which was used by 56 countries. The seven major state disinformers identified in the report are Russia, China, India, Iran, Pakistan, Saudi Arabia, and Venezuela (*The Global*, 2019).

Regardless of the relevance of Gunitsky's remarks, which democracies should take as a warning, the report clearly shows that the greatest domestic and international disinformers do not include liberal democracies. Democracies tend to use disinformation primarily for political marketing purposes, unlike authoritarian states, which use them as instruments of social control and international influence. While this book analyzes the threat of disinformation as a challenge to international security and Western democracies, subsequent parts deal with the problem from the perspective of internal state affairs.

2 The Origins of the Concept

Despite a long history of propaganda and information operations, the term "disinformation" as we understand it today originated in the Soviet Union during the interwar period and was established by the local special services. According to Ion Pacepa, the former head of the communist Romanian intelligence who defected to the West in 1978, the term and the legend surrounding it were attributed to Joseph Stalin. Stalin, he contends, used a linguistic trick to suggest the existence of a "secret science" hidden under a foreign, French name, thereby implying the wickedness and moral decay of Western countries (Pacepa & Rychlak, 2015). It can therefore be argued that the Soviet leader, who was later referred to by his own propaganda as a "great linguist", displayed his competence in this field early on by falsifying the notion that describes information falsehood in the Russian language. After all, reality is built on words.

The French themselves, however, denied the conceptual genesis of disinformation derived from their language. From the very beginning, this method of political struggle through falsehood was marked by linguistic disinformation as a deliberate misrepresentation of even the source of its origin. Interestingly, the term only appeared in Western language dictionaries over half a century later. Previously, its semantic space was exhausted by notions of "communist lies" or, more broadly, "communist propaganda". In Russia, disinformation was also elevated to the rank of a weapon used for the implementation of political strategies. It was also closely related to new concepts of conflict that blurred the boundaries between the time of war and peace, particularly when it came to offensive actions in the information domain (Wojnowski, 2015/13). There, too, significant development of disinformation methods occurred and provided a broad theoretical foundation based on the following notions:

- Active measures. These techniques refer to manipulating media and public opinion in foreign countries by using processed information and complex narratives that mix true and false content, or simply false information and disinformation. They can be disseminated using open information platforms or through the recruitment of individuals for the purpose of conducting such operations (*Soviet active measures,* 2020; Bryjka, 2019, pp. 23–38).
- Information warfare and information activities. These are understood as the conduct of information operations supporting military activities, or their independent application, to achieve political goals.
- Political war or political warfare. These concepts, seemingly absurd and tautological (because war is a political act), point to the low-level (i.e., below the threshold of classical war) use of instruments (including disinformation, propaganda, and influence operations) by states to influence the will of the enemy and elicit actions contrary to the enemy's own interests (Galeotti, 2019).

In a more modern context, the concept of intentionally concealing one's own intentions and giving specific meanings to messages is referred to as *maskirovka*, while other methods of gaining influence and controlling the behavior and reflexes of the other party are known as *reflexive control* (Thomas, 2004, pp. 237–256; Paul & Matthews, 2017; Wojnowski, 2015/12, pp. 11–36). This approach is not a one-time action but rather a continuous, long-term, and intricate influence on the opponent using prepared strings of information. The entire process utilizes narratives spread via various technical information platforms to elicit intended reactions from recipients and minimize any unwanted reactions (Kasapoglu, 2015).

Anatoly Golitsyn (2007) defined Soviet disinformation as "systematic efforts to spread false information and to falsify or block information about the actual situation and politics of the communist world" (p. 5). This definition remains relevant to this day in Russian state practice, even though the management of disinformation has shifted from the communist bloc to the Russian Federation. Golitsyn's argument on the goals and areas of disinformation practices is also interesting, including "deception", "misleading", "undermining politics", "inducing unconscious contributions to the achievement of goals", and "biased influence" on the opponent.

Golitsyn also noted that stable democratic states solve problems through open debate, which provides opportunities for disputes and increases the space for disinformation in internal communication. However, in the context of international use, disinformation is usually limited to military and intelligence matters (Golitsyn, 2007, pp. 5–8). Nonetheless, given the experience of international disinformation in the past decade, this last statement is no longer sustainable. The goals of disinformation have significantly expanded during this period, and it must now be viewed as a conscious action aimed at creating favorable conditions for political actions, with no limits to the means employed.

3 Disinformation: the Vulnerability of Democratic Societies

The complexity of creating and disseminating information, coupled with the potential of modern digital technology, has intensified the speed at which adulterated content spreads. Additionally, it has become increasingly challenging to combat disinformation systematically and to legally remove false content from circulation. The impact of a manipulated message is dependent on many variable conditions, including:

- The specific nature of the environment, which can range from open societies that allow democratic public debates to authoritarian systems that control and censor their own media space and punish those who question official narratives or disseminate "undesirable" content.

- The recipient's individual level of resistance and capacity to independently verify and process information.
- The manner and intensity of false information used in the country and abroad.
- The use of sources considered by the recipient as reliable.
- The speed and level of repetition of manipulated content.
- The speed of the reaction and defensive actions by the information relay or their recipient.

For example, the pluralistic structure of public debate, which affirms the freedom to challenge accepted beliefs and is a defining characteristic of Western societies and political systems, means that both the government and citizens rely on the same sources of information. Also, it is considered natural for both sides that messages other than their own exist in the information space. This is seen as a manifestation of the freedom of speech and diversity, which strengthens democracy and freedom. Democratic media (Oniszczuk & Gierula, 2007; Adamowski, 2008) represent various social groups and opinions and feature open ownership structures and depenalization of opinion-forming activities within the framework of legally protected freedom of speech. As a result, distorted information may appear in the system spontaneously or be placed there intentionally. However, regardless of the method of entry, such content is not only tolerated, but the political culture affirms its presence as a legitimate voice of free people and their position in public discourse.

Effective combatting of disinformation by democracies therefore becomes problematic at the level of their political and legal systems, even before the issue of detection and response techniques is addressed. This is because democracies not only encourage media pluralism and the free flow of information, but they also actively defend this area of freedom. Furthermore, such behavior enjoys social consent, which has been strengthened by decades of education in the culture of freedom, tolerance, and freedom of choice. As a result, disinformers often take advantage of the cover provided by open societies, as well as the inconsistent and profit-driven activities of commercial media operators functioning as free economic entities. This same mechanism also involves information verifiers and institutions specializing in combatting disinformation in endless and not-always-decisive discussions about the boundaries between the right to an opinion and deliberate disinformation.

This dispute also involves the challenge of distinguishing the transmitter of disinformation from its source or proving the intentions accompanying content can be classified as disinformation. The actions authorized in combating disinformation, understood in terms of security and protection of civil liberties, are often not easy to distinguish or separate from accusations concerning the restriction of media freedom or even censorship.

Even if these concerns are ignored, technical issues still arise – in the realm of effectively preventing or detecting disinformation, presenting evidence, exposing the perpetrator, and repairing the damage caused. Each phase of this process creates its own organizational and legal problems. Additionally, there is a significant delay in detecting and responding to disinformation, which increases the risk of encountering repeated instances of false information. It is also unclear who has the ability and right to combat disinformation – state authorities and services, media using technological filters of network traffic, social organizations, or activists? Another set of problems is posed by the dilemma of determining to what extent and in what areas these entities are entitled and able to operate legally, and how they should cooperate with each other.

Aside from the need for states and societies to acknowledge the inevitability of coexisting with informational anomalies, the effectiveness of combatting them should be based on two factors instead of regulations and restrictions. They are:

- Prevention. This is based on social media education, generating critical attitudes in recipients of information from an early age, encouraging habits of checking the credibility of information, and promoting comprehensive resistance to threats related to disinformation activities.
- Multi-directional disinformation response strategies and procedures. These are aimed at blocking and removing adulterated content from the infosphere, as well as correcting it. Additionally, the technology that supports disinformers can also serve to combat their activities through means such as detecting, filtering, and preventing false content, as well as utilizing machine recognition of disinformation.

Leaving aside examples of contemporary totalitarian states such as North Korea, which are based on the model of full control of the infosphere and equate a different way of thinking with the most serious of crimes, Russia and China offer different examples in this respect. Russia is currently a country with extensive experience and practice in aggressively using disinformation operations against Western countries in its foreign policy (Legucka & Kupiecki, 2022). To achieve this, it utilizes Soviet patterns or even earlier practices of the tsarist secret police. At the same time, it strongly aspires to protect its own information space. Despite the loss of its broadcasting monopoly, Moscow is still quite effective in controlling the content that appears in its infosphere, following the example of the former Soviet Union.

This is achieved through strict regulation of the ownership structures of Russian media. On the one hand, it eliminates its own "uncertain" business and civil broadcasters, and on the other hand, it limits the share of Western entities – financially, legally and administratively. The state also eagerly uses tools available under Russian law or their favorable interpretation to repress persons and civil society institutions perceived as conducting hostile information activities. Finally, control and filtering of internet content and ubiquitous state propaganda that permeates school curricula completes this picture (Giles, 2016, pp. 27–30). These tendencies, which have steadily grown since Vladimir Putin's rise to power in 2000, particularly intensified at the beginning of his presidential term in 2018, significantly limiting the inflow of external content. Consequently, they Russian public opinion was condemned to the aggressive influence of government political narratives, developed in the spirit of the official interpretation of national interests and leaving no room for independent thinking, questioning, or contestation.

This interpretation draws on the idea of the "uniqueness of the Russian soul and the fate of the nation" and tends to reinforce attitudes of respect for authority, non-alternative patriotism, and mobilization around the government's goals. It does not encourage constructive discourse with external content that presents different interpretations of facts, past events, and future scenarios (van Herpen, 2015; DiResta & Grossman, 2019). State authorities are supported in their efforts by a bureaucratic apparatus and the judiciary, which offer protection to oligarchic businesses, secret services, and numerous agents of influence. This creates a dense system of mutual support, which not only increases the chances of success but also effectively reduces the space for undesirable behavior. Those who are not convinced by state propaganda face repression or other forms of pressure. The authorities' concept of success or victory can be broadly defined, including blocking undesirable content, reducing opposition attitudes, and convincing or intimidating the unconvinced. In the international dimension, the main goal is to introduce prepared information into the consciousness of Western societies, undermining general knowledge and trust in their own sources of information, authorities, and institutions, thereby increasing susceptibility to the Russian point of view.

As previously stated, the modern world's information space is a sphere of international competition, requiring self-defense and gaining influence over the infosphere of other participants in international relations. The actions taken together shape the image of state information operations and contribute to the nature of the disinformation phenomenon. In times of war and peace, individual states are willing to rely on disinformation as an instrument of their policy in different ways. Russia and China, for instance, are an inexhaustible source of analyses and examples of its integral integration with other instruments of foreign policy. In this rivalry, modern states also exhibit varying levels of resistance to threats derived from disinformation. Generally, fewer civil liberties and more authoritarianism in governing, as well as greater media control and exclusion of foreign content, translates into a higher potential level of systemic resistance to foreign disinformation.

This assertion, however, is not a commendation of authoritarianism that uses disinformation against its own citizens. Instead, it is an attempt to highlight the vulnerability of modern democracies and societies that are open to information and cognitive threats. Disinformation is not just an unavoidable byproduct of a global network society or an objective phenomenon with deep historical or cultural roots. Rather, it should be regarded as a direct threat to democracy in all its manifestations.

CHAPTER 4

Disinformation: Anatomy and Trends

As established in the previous chapters, disinformation is a specific and developed type of message based on falsehood that is intended to paralyze the recipient or shape them in a way that works in a manner expected by the manipulator. As Vladimir Volkoff points out, disinformation has been a constant phenomenon in interpersonal communication since ancient times (Volkoff, 1999c, p. 5). When conveying information, a person can, intentionally or unknowingly, deceive the recipient by adding their own interpretation, assessment, comment, joke, or subjective opinion that is not based on an in-depth knowledge of the subject matter or any qualifications to speak on a given situation. Based on the intent of the source, researchers have distinguished three collective categories of information lies:

1. Misinformation, which does not correspond to reality; it can be spread both intentionally and unconsciously, without the intention of harming the recipient.
2. Disinformation is deliberately created or multiplied false and manipulated information; the sender's intention is to mislead the recipient for specific political, economic or military purposes.
3. Harmful information, or *malinformation,* is the misuse of real information or a stereotype in order to stigmatize specific social groups. Its includes hate speech (Lanoszka, 2019, pp. 3–4).

Misinformation could primarily be described as unintentional disinformation. It concerns a wide range of official, social, and private situations, such as the use of social media and spontaneous reactions to the content that is shared there. In such situations, the information transferred may be false, partly untrue, or unverified in terms of its truthfulness. A lack of ill intent does not, however, change the fact that even unknowingly duplicated messages can be a valuable component of a disinformation operation and be used by the operator.

Disinformation involves a form of trickery aimed at influencing the recipients of the information and effectively changing their perception of a specific issue in the direction planned by the entity carrying out the operation. In this approach, disinformation should be perceived as an element of information warfare, which is defined as "activities aimed at protecting, using, damaging, destroying information or information resources, or contradicting information in order to achieve significant benefits, a goal, or victory over the opponent" (Aleksandrowicz, 2016, p. 105). The era of digital technologies makes it

 | DOI:10.1163/9789004715769_006

particularly important to emphasize this aspect of disinformation, which uses modern methods of transmitting information to increase its range. In addition, it involves the use of fake social media accounts, false identities active on digital information platforms, or content generated automatically by bots. The scale of such influences largely defines the operational mechanisms of disinformation that function in Western societies today.

Malinformation involves the spread of false information regarding events or social groups that reinforces stereotypes, prejudices, established narratives or permanent memory patterns. This often occurs in situations related to historical facts, minority groups, or the public image of an "alien". It provides a fertile ground for politically motivated disinformation campaigns, using these established attitudes to deepen existing differences of opinion and create conditions conducive to conflict.

1 Categories of Disinformation

Despite the differing concepts and contexts surrounding the three aforementioned phenomena, each can become a component of disinformation operations conducted by states or non-state actors against third parties. Disinformation can take on a wide range of forms and can be conveyed through various channels of contemporary social communication. These channels include spoken words, conversations, jokes, rumors that are repeated and changed freely during circulation, political speeches, interviews with celebrities on television, debates featuring "pseudo-experts", manipulated films, manipulated images (including photos and graphics), sounds (e.g., music that affects the subconscious mind or song lyrics that resonate with a particular audience), written words in press articles, and text duplicated on social media.

It is important to note the informational influences within the entertainment sphere, such as satire, parody, and humorous stories, as particularly well-disguised instruments of disinformation. These influences can act on the recipient irrespective of their intellect and knowledge, using their abstract thinking abilities or sense of humor. It is often said that authoritarian regimes do not have a sense of humor but are happy to use it offensively in information operations. They try to mock, caricaturize, or otherwise depreciate and delegitimize certain figures, groups, or institutions in the eyes of the recipients, unable to attack them directly. It can also be particularly easy to cast or use unaware artists, tabloid journalists, performers, and internet or TV entertainment influencers in this role. Lowering the recipient's sensitivity to this type of manipulation and misleading is favored by the general decline in the quality of this type of production, which has naturally increased its influence during electoral campaigns, referenda, and other important public decisions. This does not mean that open societies should be afraid of their own sense of humor and entertainment. However, it is essential for their media education to teach them to be sensitive to this channel used for disseminating disinformation.

The list below summarizes the 12 possible forms of communication that may contain disinformation. The items mentioned most often appear in professional analyses of this phenomenon presented in relevant literature. This list is not comprehensive, and it is worth remembering that each of its components may appear differently depending on the specific medium it uses.

1. **Satire and parody** are acts that serve to discredit the object of disinformation, either intentionally to cause harm or unintentionally with the potential to stun the recipient (Rashkin et al., 2017, p. 3).
2. **Rumors** involve a change of the initial truth elements found in information during its circulation.
3. **Deceptive content** is the specific use of information to present a fact or person in a specific light or to authenticate a fabricated situation or belief based on unfounded messages.
4. **False content** is content that pretends to be original.
5. **Forged content** is intentionally and fully falsified information used with the purpose of deceiving the recipient.
6. **False connections** occur when a text, its title, or an image is used but does not reflect its actual content or meaning.
7. **False context** involves the use of true content conveyed within a false context.
8. **Manipulated content** is when original content is distorted to deceive the recipient (Wardle & Derakhshan, 2017, p. 17).
9. **Myth** a repeated message containing unverified and intricate information. In foreign policy, states can create myths using multiple narratives, manipulated in terms of the accuracy of facts and the way they are presented. These myths may evolve over time and be altered or updated based on the requirements of the creator as well as the changing circumstances (Kupiecki, 2019, pp. 77–105).
10. **Propaganda** is content aimed at influencing people's minds for a specific intended effect.
11. **Sponsored information** features a hidden form of persuasion conveyed as objective information.
12. **Information error.**

European Union analysts (*Modus trollerandi,* 2021) have revealed the empirical mechanism of interaction between various types and techniques of disinformation in the pursuit of offensive political goals. Although the perpetrators may appear to accept the standard of open debate, their actions are aggressive and aimed at narrowing the communication space. They seek to eliminate not only rational voices but also all fact-based positions that differ from their own. Disinformation can therefore be, to some extent, considered a model of modern information warfare that is based on the use of seven combined tactics:

1. Attacking views never expressed by the target of the interactions.
2. Spreading *whataboutism* and deflecting the discussion away from the original topic.
3. Using offensive or inflammatory language to discourage opponents.
4. Mocking and using sarcasm to minimize the voice of opponents.
5. Provoking and asking who benefits from it.
6. Bringing opponents to exhaustion by drowning them in detail and technicalities.
7. Strongly denying any evidence of disinformation (*Modus trollerandi,* 2021).

The mechanism described above refers to the correlation of three rhetorical techniques known in linguistic studies. Ethos defines a goal (as a person or a group), pathos affects emotions, and logos operates in the area of weaknesses of logical processes (Sample et al., 2020, p. 8). The same team that developed *Modus trollerandi* quotes a different seven-step scale from the *New York Times* that denotes the actions used by disinformers, which coincides in some places with the above-mentioned analysis. It goes as follows: look for divisions; create a lie; wrap the lie in the appearance of facts; hide your share; find a useful idiot; deny everything; and play the game long enough (*Seven commandments*, 2018).

2 The Building Blocks of Disinformation

Disinformation agents target their opponents' weaknesses and aim to change their point of view rather than engage in respectful communication and exchange of views. They perpetrate acts of informational violence, attacking the opponent's cognitive mechanisms and exhausting them with a variety of continuous interactions, ranging from jokes to offensive attacks, all under the guise of pseudo-rationality. The form of disinformation can be deceptively innocent as it is often intertwined with its carrier. Moreover, any form of manipulated information can take on a life of its own and create a story whose origins and destiny can only be understood by exploring the larger narrative of which it is a part (Bal, 2009; Ricks, 2015).

Experts point to relationships between three formally separate building blocks:

- A given story or narrative. This is about showing specific events in a specific context.
- A specific message. This can present in the form of an image, text, sound, or a combination.
- A plot or meta-narrative. This plot plays a superior role compared to the messages related to it and organizes the variable messages, giving them an axiologically unambiguous character on the axis of good versus evil or in a clear hierarchical order.

The message, plot, and narratives can create an infinite number of versions of the presented reality, known as plot modifications. These modifications arise depending on the specificity of the recipients, the expected effects they are intended to cause, and the effectiveness of the chosen form of communication. The story itself is therefore a function of the mission it fulfills in the information operation. Its content, emotional charge, and the percentage of truth contained in it constitute only an operationally useful, time and place-specific mutation of a strategic meta-story that transcends time and exists beyond its tactical carriers and records (Kołodziej, 2017, p. 26). Such a message serves as an auxiliary to the key goals of politics and has only as much value for it as the result it is capable of producing. It lives only as long as it remains effective and can be freely changed to ensure continued correlation with the overarching assumption it serves.

In the practice of Russian disinformation, there are many such duplicated plots using fabricated meta-stories. Examples include:

- The "betrayal of the West", cited as the cause of Russia's aggressive actions in the world and presented as a justification for enforcing the right to self-defense. This narrative is employed in operations directed against Western countries as an offensive argument introduced into the local political discourse, as well as in operations directed towards its own citizens for the purposes of social mobilization and consolidation around state leadership.
- The narrative that frames Russia's actions as a defense against the imposition of Western values and way of life, which are presented as degenerate and alien to the "Russian soul". In political terms, this is manifested by the West's support for pro-democratic forces and movements in the countries of the former Soviet Union, commonly referred to as color revolutions.
- Narratives around Western efforts to colonize Russia, use its natural resources, and employ Russians as cheap labor.

Narratives based on such plots are disseminated to Russians and the international community through various channels, often relying on what is known as the “big lie”. These narratives not only violate historical facts and truth but also common sense (Houston et al. 2015). To better understand how such stories are built and disseminated, the study of lies in journalistic texts can be particularly useful. The findings from these can be applied to lies presented in both traditional and new media information messages. A Russian media expert from the University of Kazan has listed nine types of these forms:

1. A paranoid lie. This involves presenting completely unverifiable facts, with an argumentation system that is highly ideologized and does not reflect rational understanding. Such messaging strongly affects the recipient’s emotions and often uses the image of an enemy who is blamed for all the evils or conspiracies surrounding the subject, who presents themselves as the victim.
2. A politically convenient lie. The factual basis of this message is often limited, and facts are chosen selectively, mixing truth and falsehood. This can disturb the recipient’s understanding as they may assume that the described events are probably factual.
3. A discrediting lie. This message is masked by ambiguous concepts or complicated images.
4. An embarrassing or demagogic lie. The aim of this message is to discourage the recipient from engaging with the described matter or person permanently by using general and oversimplified statements, creating scapegoats and mental shortcuts, and portraying negative characters or features.
5. An interpretative lie. In this type of message, facts are sparsely used but deliberately thrown around in a way that leads the recipient to create a specific story. The narrative may include cleverly expressed opinions that influence the recipient’s perception and interpretation of the facts.
6. A persistent lie. This is a message in which each utterance contains or refers to previously known narratives.
7. A conceptual lie. This type of falsehood is linked to the prevailing ideology, and altering it can result in the authentication of narratives about the past.
8. A hypostasized lie. This message imparts the actual definition of an abstract concept, property, or idea, and has a subliminal effect that emotionally inspires the recipient toward a particular action. The liar in question eagerly employs libertarian references and the subjective nature of the opinions.
9. An axiological lie. Arising from the assessment of values that favor our own (i.e., “our values are superior to those of others”), such evaluations supplant logical reasoning, gaining a persuasive character and the strength of arguments (Ostaszewski, 2018, pp. 36–39).

Once again, the above categorization of media lies demonstrates that they need not be entirely false to serve as instruments of influence. It suffices to selectively choose and "adjust" facts and to skillfully interpret and package them linguistically or visually, placing the recipient in a situation where they face an "apparent" choice of a position dictated by the "heart" or one that purportedly stems from proper cognitive processes, all orchestrated by the disinformation source.

3 Trends in Disinformation

In the third decade of the 21st century, disinformation is thriving and its prospects are impressive. This is due to the anticipated demand for social influence tools in the realm of politics, including elections, party rivalries, personal campaigns by politicians, relations with society, legitimacy protection, and governance systems. Similar trends exist in marketing and market competition, where disinformation is akin to doping in sports. The data cited in the previous section demonstrates that one-third of countries, or the political and economic entities within them, employ varying degrees of disinformation tools. While the context and purpose of this practice as well as its relationship to foreign policy are diverse, its potential for expansion is significant. These needs will continue to grow in the international arena as actors seek to enhance their power at the expense of others, using disinformation as a tool. This trend is also linked to the evolution of peacetime conflicts, where finding effective methods of influencing the decisions and cognitive processes of opponents has become important, beyond the traditional measures of military potential and other "hard" policy instruments. The widespread use of disinformation and its social standing on par with truth poses a global threat to individuals, societies, and states and their military bases, economies, and political systems. According to Wardle and Derakhshan (2017), three aspects should be considered in the developmental trends of disinformation:

1. The learning processes of disinformers and the lessons they draw from the past to improve their tools and activities.
2. The development of technology, which has expanded the boundaries of imagination regarding effective information interactions and their social consequences. If there is any reason to be optimistic about the impact of the disruptive digital technologies, it is connected to their "dual use" nature (Kupiecki, 2020a, pp. 472–497). While these technologies allow for even more effective and pervasive disinformation, they also provide an opportunity for these same technical capabilities to be used for education, detecting, and combating information threats, and strengthening

the resilience and cognitive security of modern societies. In this context, *cognitive security* is a relatively new area of research related to security. It pertains to activities aimed at defending societies against the harmful effects of disinformation and information manipulation supported by modern technologies.
3. A relatively new but fast-learning actor in the field of disinformation: China (Wardle & Derakhshan, 2017).

Scientific analyses of disinformation, regardless of the research discipline in which they are conducted, tend to overlook important aspect of this problem, namely the *post-factum processing* of experiences by information operations strategists. This analytical phase involves estimating the costs and benefits of a given operation, evaluating the effectiveness of the techniques used, and assessing the evolution of the operating environment. It is an integral part of planning similar activities in the future and serves as a process of continuous learning to improve information interaction methods and develop the repertoire of analysts' applications in various situational contexts. A good example of disinformation operations as a learning process is the Russian influence on the electoral processes in the U.S. in 2016 and 2020.

According to an expert from the Helsinki Center of Excellence who deals with hybrid threats, the initial stages of similar campaigns involved the placement of harmful content in the enemy's information space directly by the Russian entities conducting the operation, located outside the US. In subsequent years, the Russian *modus operandi* in this area changed. They reduced the risk of detection and the connection with Moscow by acquiring American news portals that spread the desired narratives for their own activities. These narratives focused on issues that polarized American public opinion (e.g., the myth that the 2020 U.S. election was stolen from Donald Trump by the Democratic Party). The websites were run by genuine American activists operating on platforms such as Twitter, YouTube, and Facebook. The content they disseminated was difficult to unequivocally classify as false but it was prepared in such a way that its messages would strengthen attitudes of dissatisfaction and protest, delegitimizing the new U.S. president. At the same time, these activities emphasized the need to reduce the amount of machine-disseminated information in favor of people who could be identified without suspicion as aspiring opinion leaders or "concerned citizens", including influencers with millions of followers.

At the same time, many different types of media are used to disseminate information so that no administrative action (e.g., the removal of false content) can disrupt the entire operation and weaken the influence of a given message. During the 2020 American elections, the disinformation employed was primarily emotional rather than based on knowledge. Emotionally charged content from social media was disseminated on television news programs and popular talk shows through many message sources. This made it difficult to detect and effectively counteract false information in the face of the dynamic nature of the messages and the confusion between truth and falsehood within the emotional content.

The means and methods of disinformation are constantly evolving and adapting to changing conditions, including the circumvention of defense mechanisms. Perpetrators can better conceal their identities behind fake social media accounts or pay local internet users to spread disinformation. Detecting disinformation in podcasts (Wirtschafter, 2021) or even audio disinformation (Urbani, 2021) has also become more time-consuming and challenging, especially as this type of message becomes more popular and difficult to verify through online moderation. Trolls working for Russia on social networking sites often leave comments under articles on mainstream Western media portals, which are then used in pro-Kremlin propaganda as "favorable voices" from the West. This new approach has been further complicated as major social media platforms increase their efforts to counter disinformation (Corera, 2021).

Disinformation as a global problem has manifested itself in various ways, such as attacks on Australia from the Balkans and Israel, or the organization of anti-vaccine protests from Germany in the Antipodes' (Galloway, 2021). Ghana's cheap English-speaking disinformation operators, for instance, have been hired to interfere with the U.S. election process. During the 2020 U.S. presidential elections, troll groups were created in Ghana to pose as African Americans. Low-cost account operators from the Balkans, such as North Macedonia and Kosovo, were also hired to pose as "voters" struggling online, potentially reaching 140 million Americans a month. In Kenya, a disinformation operator for hire is valued at only $15 per day (Elliott, 2021a). Fake social media accounts created on behalf of Russian services and their agents covered Facebook pages of American Christians, African Americans, and Native Americans. Additionally, 70% of online racist attacks on dark-skinned English footballers come from abroad (Stokel-Walker, 2021a).

Conclusions from disinformation campaigns to date can be formulated in three ways.

First, they can be seen through the paradox of massive dissemination of false content that is perpetrated by genuine users. This has guaranteed the speediness of information distribution, making it impossible to control by portal administrators and fact-checking organizations.

Second, there is an increase in the number of partially falsified messages compared to completely fabricated ones. This technique is a reaction to the actions of internet portals filtering election content and hate speech. Manipulated or distorted information is more difficult to detect using control algorithms.

Third, there is a cascade of network interactions not related to one medium or digital platform but simultaneously using many different places and tools of communication, including encrypted messaging (Garcia-Camargo & Bradshaw, 2021).

4 Technological Challenges

Modern technological advancements in various fields have affected both the development of disinformation and the ability to combat it. Digitization has become prevalent in all areas of life, including work, services, entertainment, and communication. Yet the most significant revolution is happening in the field of artificial intelligence (AI) research and its applications. Many believe that AI, with its enormous data sets, is already demonstrating its capabilities and will pose a greater threat to information security, privacy, database integrity, and systems for processing and sharing knowledge in the future.

The problem is not only the faster spread of false information in greater quantities but also the increasing use of human identity as a tool for disinformation. Techniques like *deepfakes*, which can manipulate images, voices, locations, intentions, outputs, and reputations, pose a criminal risk when used to infringe on personal rights. Much more serious problems may arise in the political sphere, however, when the victim of such actions becomes a state leader, for example uttering content that delegitimizes their own leadership or poses a threat to peace. The capabilities of AI in this area go beyond entertainment applications and will have far-reaching implications for human cognitive processes, education, everyday life, and international relations, with increasing application in both the civil and military arenas.

Even today, there is a real technological "arms race" taking place among world powers that recognize the potential of new technical solutions and the expansion of their applications. Countries that gain an advantage over others in this respect or dominate development processes will strengthen their international position, defined today by traditional determinants of power such as army, territory, and capital. The essence of this matter has been laconically and precisely expressed by Vladimir Putin, who has asserted that the first country to master artificial intelligence will dominate international relations (Vincent, 2017). The ambitions of the Russian president and China's leader in this regard are prompting Western leadership to seriously reflect on their own strategies.

The West is already experiencing aggressive information operations that utilize the developing functionalities of artificial intelligence, which includes supporting social communication processes as an "accelerator" of the transmission, individualization, and multiplication of information. AI's role as an autonomous content creator heralds new though still unavailable opportunities for disinformation on a massive scale as a tool for states, corporations, and criminal organizations. The modern world's dependence on information and its internet transmitters, combined with the possibilities of machine learning, creates a real explosive mixture. The actual aftermath of this mixture will only be revealed in the future, however. Just like dynamite or firearms, which have become tools of both destruction and development, artificial intelligence may become the greatest threat to or source of protection for the information security of democratic societies (*The phenomenon of disinformation*, 2019).

The task of serving the truth is not simple, however. It is determined not only by the scale of technological challenges but also by the legal, organizational, political, and cultural conditions that make up the social environment of artificial intelligence applications. Paradoxically, challenges for defenders of the truth outweigh those of the producers of disinformation, who do not care about truth, universal values, or social order. In democratic countries, determining what can be considered truth and what should be fought as disinformation begins with questions about the subject and its legal basis. Justice institutions and state authorities, civil society organizations, and people's habituation to freedom of choice, guard the appropriate justification of the answer. Arbitrary administrative actions by governments and media owners, for instance, would face charges of censorship and restriction of freedom of speech.

Another problem is the complex regulatory context of information governance. This includes methods of controlling the use of modern technologies or algorithms in their specific applications to prevent discrimination against specific groups. Another issue is ensuring fair cooperation between governments, the European Union, and producers and users of information. Governments act as regulators interested in maintaining order, while media owners are interested in profit and avoiding regulation. Creative employees and the audience of their works are usually reluctant to limit their freedom and are prone to protest if the actions of regulators arouse their suspicions.

There is a growing awareness among all these groups about the scale and dangers of online disinformation. Artificial intelligence could be a possible solution to limit these dangers as it operates systematically and faster than human verifiers of information. Social media owners, under pressure from governments and conscious users, have long been investing in algorithms detecting *fake news* and removing false content created by bots from circulation. They have also been collaborating more and more with non-governmental organizations to combat disinformation using the capabilities of AI. Although the elimination of bots through the analysis of network traffic and patterns of undesirable behavior supports the fight against disinformation, it does not solve the core of the problem that bots present. Bots operate based on how they have been programmed by humans and therefore possess no remorse for the damage caused by the misinformation they spread over the internet.

Analyses and experiments conducted by specialists show that bots duplicate both truth and falsehood equally; it is humans who are predominantly responsible for the deliberate or accidental spread of disinformation. Effective media education, critical thinking, and forming habits of fact-checking are therefore of great importance. Combatting disinformation must rely on both the power of computers and human presence of mind to pay attention to distortions or inconsistencies in images and content. There are many widely available and rapidly growing *fact-checking organizations* in the world based on cooperation between humans and machines. For example, companies verify text, image data, and information flows on social media using AI algorithms through their applications. More and more often they can also filter toxic content such as hate speech or obscene comments.

Given the current state of artificial intelligence, it may not be realistic to expect fully effective tools in the near future to combat internet disinformation. However, it is worth considering AI as a solution that can support people who act ethically and in accordance with safety requirements. The future of AI in relation to disinformation should therefore be viewed as a tool with dual purposes rather than an inevitable threat to be feared. Depending on human decisions, it can be wielded as a sword by opponents of freedom or used as a shield to protect societies and individuals from the effects of disinformation. This comparison, often used in history by military strategists, applies to a complex modern battlefield where much still depends on human actions. The choice is clear.

5 China: the New Actor of Disinformation

China's emergence as a major player in the realm of international disinformation is a recent development that can be attributed to its foreign policy reorientation in the 21st century. The Chinese government's overt expressions of superpower aspirations, regional leadership, and efforts to create favorable conditions for its own interests worldwide have led to an open, multidimensional rivalry with the United States and the Western world as a whole. This rivalry spans across various spheres, such as politics, the military, and the economy, with key focal points including the technological arms race and the expansion of China's "soft power" components. Against this background, information and information operations, which are deeply ingrained in Chinese strategic thinking, are viewed as a tool to improve global conditions and help implement state interests. As a component of military operations, they also have specific uses, which were discussed in Chapter Two.

In the opinion of a U.S. Atlantic Council expert:

> In recent years, China has increased the activity of its own state propaganda and manipulative actions in social media to promote its own vision of the world towards the people of Hong Kong and Taiwan and the Chinese diaspora in the U.S. and other countries. Aggressive promotion of one's own vision of the world coincides with the growing one economic, political, and military power of this country. (Roberts, 2020, pp. 6–10)

China has a long history of propaganda, which after the victory of the Communist Revolution in 1949 became a tool of both social communication and government oppression by the regime in Beijing. For decades it focused on its own society and its neighbor in Taiwan. Although both these targets remain high on the list of contemporary Chinese information operations, in the dimension of international disinformation China is emerging as a new actor with growing influence. It is drawing on Russian models while developing its own methods of influence using traditional media and direct messages phrased by state functionaries. It is also setting new directions for disinformation using social media, bots and artificial intelligence.

China spreads disinformation using both state media and global social media platforms. Their messaging centers on issues of sovereignty and non-interference in the internal affairs of the "Chinese world". They also strongly exploit the image of a state that was humiliated in the past by Western powers, which they claim gives China the right to defend its ownership status and decide on its directions of development.

The complex crisis surrounding the COVID-19 pandemic has led to an unprecedented collaboration between Russia and China in their international communication strategies, which has resulted in the strengthening of propaganda and media cooperation on both sides (Legucka & Przychodniak, 2020). In this partnership, Russia gained an ally in its opportunistic disinformation campaign, and China found a defender against accusations of being responsible for the global viral crisis (*China on*, 2020). This tactical alliance demonstrates that the foreign policy objectives of both states, which seek to revise the liberal international order, are similar. They both feel discriminated against, have limited respect for its rules, and are willing to change or expand the freedom of interpretation of these rules (i.e., the rules of international law). However, they differ in the means they use to achieve these objectives.

Russia primarily uses political corruption and military pressure against foreign countries. China, while consistently expanding its military potential in Asia and its power projection capabilities, emphasizes economic pressure measures. In the field of information, China is in fact a country closed to external influences. It strictly control its own information space and censor the internet and the largest web search engines. The state's information monopoly is protected by the ownership structure of the media and by laws, which penalize the dissemination of content that goes against the messages of local authorities. It can therefore be said that, in a way, Beijing has created an endemic version of social resistance to hostile influence. At its essence, however, is not the freedom of choice and education but control over the population and limitations of individual freedoms and information rights. Despite this strict control, the state and its institutions fully enjoy the benefits of globalization and an open economy. China's ambition to re-write the history of the 2020 pandemic that ignores its origins in Wuhan has a much longer-term significance. It is aimed not only at rebuilding Beijing's tarnished international image but also its "soft power", which it uses as an instrument for fulfilling its political and economic interests.

While there are similarities in Russia's and China's goals and methods of disinformation, it is important for observers, analysts, and communication practitioners to recognize significant differences in their current and long-term goals, as well as their methods of achieving them. Russia has been responsible for most disinformation and media influence operations worldwide, while China has recently begun replicating similar patterns of aggressive disinformation. The sophistication of China's disinformation techniques is also growing rapidly, aided by advances in technology, including artificial intelligence, which will play an increasingly important role in enhancing the precision and responsiveness of Chinese information operations.

Additionally, China's economic relations with foreign partners, particularly in accessing the Chinese market and cooperating in "soft" areas such as film production, will increasingly link access to concessions and loans with the neutrality of foreign partners or their active promotion of the Chinese vision of the world and its image as a stable and peaceful superpower. Chinese disinformation seeks to promote its own efficiency and agency against the backdrop of a supposedly ineffective rotten West, both domestically and internationally. Beijing will support this narrative through information interactions and broader influence operations that are continually refined.

Both experienced disinformation operators utilizing older schemes and methods and novices find their place and application in the new reality. One example of this is seen through the Chinese conspiracy theory claiming that the U.S. military garrison of Fort Detrick is responsible for the origin of COVID-19 (*Wuhan*, 2021). This theory follows similar patterns as the old Soviet propaganda that accused the U.S. of creating HIV, as well as recent Russian myths about American laboratories experimenting with biological weapons in Georgia or Ukraine. So far, the russification of Chinese disinformation efforts has progressed less effectively and without finesse.

This also seems to be the case with French-language Chinese state media portals, which have large followings but record disproportionately low engagement, with their political propaganda in cultural content appearing clumsy. Nevertheless, China has been allocating increasingly significant resources to coordinated disinformation campaigns worldwide, with an annual spending of around EUR 1.3 billion on information operations abroad even before the pandemic (*French-Language*, 2020). Moreover, with the continuous improvement of data processing ability, it is now possible to distort reality even faster, more perfidiously, and in a way that is more difficult to detect. For instance, a network of 350 fake accounts spreading pro-China narratives was identified, and their profiles were found to have been created with the help of artificial intelligence. These accounts focused on narratives related to COVID-19 and problems with respecting human rights in the U.S., including the murder of George Floyd during a police arrest. The mesh was detected partly by analyzing hashtag usage (Carmichael, 2021).

The Chinese activity behind disinformation is largely driven by the accelerated effect of COVID-19 and often inept attempts to imitate Putin's trolls. However, it must be acknowledged that China is currently in a transitional stage, in a period of apprenticeship after which it will create its own school of international disinformation, harnessing the results of its research on artificial intelligence in a much more effective way than before. To carry out these activities, China will use the potential of Chinese diasporas around the world, as exemplified by the ongoing problems in Australia, Canada, and the U.S. The scale, tools, and perspectives of multifaceted operations by the People's Republic of China are illustrated by an excellent 650-page study prepared by analysts at the Institut de Recherche Stratégique of the École Militaire in Paris (Charon & Jeangene-Vilmer, 2021).

Summary

1. What is disinformation?

Disinformation is both a doctrine and a practice employed by states or non-state actors to deliberately use manipulated or falsified information to induce a desired change in a specific group of recipients in the targeted field of influence. It is often used as part of propaganda operations that utilize techniques of influence and psychological manipulation during times of peace, crisis, and war in order to harm the targeted audience.

2. What is international disinformation?

International disinformation refers to the deliberate use of disinformation techniques and tools by states or other participants in international relations as a means of influence in foreign policy or other international activities during times of war and peace. Its goal is to manipulate the recipient of the information and improve the sender's situation. Disinformation activities are usually targeted at foreign societies or large groups within them.

3. Is disinformation a product of the 20th century?

Although the concept itself was developed in the Soviet Union in the 20th century, the phenomenon is as old as human history. As a tool of war and politics, disinformation has been used since ancient times and is referenced in historical writings and the oldest treatises on strategy.

4. Is disinformation a threat?

Disinformation damages social trust and undermines the values, institutions, and political processes of democratic nations. It erodes the foundations of good governance and limits the ability of people to make rational decisions, which rely on the provision of truthful and credible information.

5. Is disinformation always based on false information?

It can use both fabricated false information and manipulated information that contains elements of truth and falsehood. The intention of pure disinformation is to mislead the recipient, regardless of the tool used or the content of the message.

6. Are disinformation and propaganda different names for the same phenomenon?

Although propaganda and disinformation are similar in that they both use information messages, they are different instruments of state policy.

Propaganda, depending on its type, is not necessarily intended to harm the recipient. It can, in some iterations, present as a message that is transparent about its goals and the identity of its sender.

7. What is fake news?

Fake news is a type of disinformation that involves the dissemination of fabricated or partially true information in a manner that imitates reliable information. This can take many forms, including distorted text, images, sound, or videos, and can even include the creation of *deep fakes* featuring public figures.

8. Where does disinformation most often occur?

Disinformation can be disseminated through various means, including direct communication between people, traditional media, and modern internet platforms and social media. With the advancement of digital technologies, internet-based media has become a particularly frequent carrier of disinformation in recent years.

9. What forms can disinformation manifest in?

Disinformation can take many forms, including falsehoods hidden in forms that reduce the recipient's vigilance. One form is satire and parody, used to discredit authorities or distract the recipient. Others include false content from a seemingly credible source, false texts related to true information, false relationships conveyed by texts or images with distracting titles and comments, or myths based on repeated messages that contain information not supported by evidence.

10. How do I protect myself from disinformation?

It is important to verify both the content and the source of information before accepting it as true. One should not react impulsively based on emotions or blindly share unverified information on social media. It is advisable to seek out reliable sources of information and engage in discussion with others to compare and contrast different perspectives.

PART 2

Disinformation: Recognition and Analysis

CHAPTER 5

Intelligence and Military Disinformation and Its Impact on Information Security: Theoretical, Practical, and Legal Aspects

We live in a network society where a significant part of our personal and professional activities take place in cyberspace (Castells, 2011). The rapid socio-economic changes and technological advancements in recent decades have transformed the way we function in our everyday lives, communicate, acquire knowledge, and perceive the world. The internet has made it possible to shop, learn, work, make friends, and communicate without limitations of time zones or distance, and it allows us to stay updated on events from around the world. Churches conduct their missionary activities, terrorists recruit fighters, and states wage their conflicts – all online.

This widespread use of the internet has caused many users to lose their sense of critical distance, self-control, and distrust toward the content published on it. The situation has created opportunities for foreign intelligence services to conduct disinformation operations that target decision-makers and entire societies. Furthermore, the information and technology revolution has altered the nature of military information warfare at the strategic, operational, and tactical levels. State institutions, armed forces, and societies must therefore develop effective defense mechanisms to combat falsehood and manipulation.

1 State Information Security

In today's world, information is considered a strategic asset, whether it is viewed from the perspective of an individual, society, the state, or the international community. Access to information and the ability to process, transmit, secure, and store it determines cognitive security. These conditions are essential for effective decision-making, management, and operation of all entities, including states, institutional systems, private sector entities, societies, and individuals. In contemporary times, knowledge and information technologies have become one of the primary factors of production. It is estimated that the IT sector is already responsible for creating over 15% of global economic growth, showing an upward trend (Henry-Nickie et al., 2019).

In the 21st century, the world's economy and global politics are heavily reliant on information, its accuracy, timing, and uninterrupted flow. The consequences of long-term disconnection from telecommunication networks, also known as internet blackouts, could be severe. Such a situation would bring about the paralysis of most employment sectors, stock market crashes, stop in financial flows, and disruption of the functioning of critical infrastructure that sustains vital areas of state responsibility. Inevitably, it could lead to social unrest, chaos, riots, bankruptcies, and even the collapse of governments.

Because of this, it is hardly surprising that in reflections on the future of international conflicts, cyber threats and offensive network scenarios are now more seriously considered than traditional military clashes. In 2015, as a result of internet blockades, the global economy suffered losses amounting to USD 2.4 billion (Waddell, 2016). In 2019, this amount grew to USD 8 billion before dropping to USD 5.5 billion in 2021 (Woodhams & Migliano, 2021).

From the state's perspective, information, its accessibility, and the possibility of safe processing are critical components of its power. It is also a crucial element for its functioning in politics, security and defense, the economy, as well as in the social and cultural dimension. Sensitive information containing data that requires special protection against unauthorized access must therefore be adequately protected from the moment of acquisition, during processing (analysis), transfer, storage, and usage, with appropriate standards implemented at each stage. State information security is ensured when:

- Information resources are not at risk.
- State institutions make decisions on the basis of true, relevant and accurate information. This does not guarantee the quality of the decisions themselves but provides conditions for rational and optimal actions.
- Information flows and the functioning of the information and communication technology (ICT) networks that make up the state's critical infrastructure are undisturbed.
- State structures effectively ensure the protection of classified information.
- Public institutions do not violate citizens' right to privacy and the protection of their personal data.
- Citizens, non-governmental organizations (NGOs), and the media have access to public information.
- Society is resistant to disinformation and propaganda activities (Aleksandrowicz, 2018, pp. 33–35).

Information security is a trans-sectoral area that encompasses the information environment, including the state's cyberspace. A threat arises when state structures are unable to provide effective protection, making society susceptible to falsehood and manipulation. The core of the problem is securing the functioning of the state (and social order) and protecting its interests in the information space.

The above approach to the problem emphasizes three issues:

1. Protection of the state's information resources. This particularly applies to protecting sensitive and classified information that is crucial for its functioning and of strategic importance against unauthorized access (e.g., espionage). This also requires protection against disruptions of its functioning due to cyber-attacks, acts of sabotage, etc.
2. Protection of state institutions and society against the impact of disinformation and propaganda. These disinformation activities may be aimed at causing social unrest or internal destabilization of the state; interfering in political processes like elections; or shaping internal, foreign, or security and defense policies.
3. Retaining offensive capabilities against the information resources of potential opponents. This involves pursuing tools to be able to influence their societies in an informative way.

To protect sensitive information against unauthorized access, state institutions impose secrecy clauses on it. These commonly fall into classifications such as "restricted", "confidential", "secret", and "top secret". NATO also has classification categories indicating the originator/owner or nature of the protected information, with symbols such as "cosmic top secret" or "atomal top secret". The content of information protected by these categories is strictly regulated by law, and administrative proceedings are instituted for persons applying for access to them. Access is granted through a limited, strictly personal admission to a certain level of sensitive data, which does not entail automatic access to all such information. To ensure protection, states restrict access to only the group of people who need the information to perform specific task; they also increase the physical and technical security of their processing systems.

They do this through complex and detailed office systems, which include conditions for storing a given type of knowledge, accessing it, and learning about it. This takes place, for instance, in government institutions, the armed forces, the defense sector, the arms industry, the organs and services of the state security system, public administration, and other elements of critical infrastructure. Specialized counterintelligence institutions, referred to in NATO terminology as National Security Authorities, are responsible for guarding state secrets and preventing their unauthorized disclosure to unauthorized persons. The tasks of these institutions include:

- Controlling classified information and ensuring compliance with the applicable rules and regulations.

- Securing ICT systems of institutions that have access to classified information.
- Conducting verification and control procedures as well as proceedings within the scope of industrial security.
- Ensuring the protection of the exchange of classified information with other states and international organizations (e.g., within NATO and the EU).
- Providing advice and training in the field of classified information protection.

Access to classified information, or a security clearance, is only granted to trusted and authorized persons who have undergone complex vetting proceedings confirmed by the National Security Authority. Classified information must be processed under conditions that prevent its unauthorized disclosure, and this is regulated by relevant provisions specifying the requirements for secret offices, ICT system security, material circulation, and physical security measures. The "need-to-know rule" states that classified information is made available only to authorized persons to the extent necessary for the performance of their official duties. This means that not everyone with a security clearance up to the "secret" level has the right to access all information within that classification. The principle of adequate protection of information also applies to the person who comes into possession of it. Institutions and companies with industrial access to classified information are subject to control regarding how they protect it, and counterintelligence services work closely with the divisions responsible for protecting classified information in individual institutions. The unauthorized disclosure or use of such information is a punishable crime.

2 Intelligence Disinformation

Against this background, it is worth noting the category of "intelligence disinformation" as a special type of manipulation of cognitive processes and international communication. Although similar to "classical" disinformation, its differentiation lies in the specialized nature of the institutions carrying out such activities, as well as their organization and goals. Intelligence disinformation can be defined as "the process of influencing the behavior of a disinformed subject by distorting their perception of reality, leading to taking actions consistent with the deformed image, and at the same time corresponding to the interests of the disinforming entity" (Świerczek, 2020, p. 33).

By delivering prepared information to the opponent, foreign intelligence services aim to make the target of disinformation believe in the credibility of the information provided to them, leading them to draw conclusions and make decisions that are inconsistent with the actual state of affairs but consistent with the goals of the entity conducting the operation. These decisions may result in actions that are detrimental to the interests of the disinformed entity, such as improper allocation of resources and forces, errors in the assessment of threats or opponent intentions, or a false sense of security, such as when the aggressor misinforms about its hostile intentions. The purpose of intelligence disinformation may involve:

- The falsification of knowledge about the state's military, diplomatic, intelligence, or economic capabilities.
- Concealing the actual strength and integrity of the state's subsystems, as well as its strategies and intentions.
- Ensuring the political, social, and economic stability of the disinforming entity and at the same time concealing problems. This could include economic issues, internal disputes, factional struggles, crises, or social unrest.
- Influencing the implementation of the policy of a disinformed state (e.g., in the area of security and defense or allied policy).
- Building foundations for further psychological and disinformation operations.
- Distorting real information in such a way that it becomes useless for decision-makers.

When conducting disinformation operations, secret services engage in deliberate, planned, and covert actions with assistance from various entities that use different channels of information transmission but are subject to the supervision of a centralized directing center. Disinformation is not only the domain of intelligence services; it also used by counterintelligence services as part of their 1) active counterintelligence operations, which involve disinformation operations; 2) offensive counterintelligence, which is carried out with the help of a network of double agents. Disinformation is the primary operational tool in both cases.

Intelligence and counterintelligence disinformation activities, carried out in a specific environment with set goals, usually consist of the following stages:

- Selection of channels for transmitting disinformation. These can include political and military elites, state-owned companies, an analytical and expert base, journalists and commentators, influence agencies, "useful idiots", trolls, and bots.
- Determination and possible elimination of alternative sources of obtaining and verifying information by the targets.

- Masking of actions that lead to operational initiatives. Maintaining secrecy ensures that the disinformed object believes their decisions, successes and failures are the result of their own actions or mistakes only and not external influences (Hosaka, 2020).

Politicians, business people, government officials, journalists, bloggers, vloggers, commentators, experts or individuals who use a fabricated identity can act as inspirers or transmitters of disinformation. They may utilize their accounts on popular social media platforms such as Facebook, Twitter, YouTube, Instagram, Tik Tok, or Russian vKontakte and Telegram. Another category of relays and information producers on the internet are "machine bots", which are computer algorithms that automatically create and duplicate false information, posts, comments, or other content after appropriate programming.

"Agents of influence" are individuals recruited by foreign intelligence and consciously acting on its behalf. They follow instructions for which they may receive specific benefits (e.g., financial); their activities can also be ideologically motivated or the result of blackmail or psychological and personality traits. The motivational model used by secret services to obtain human assets is commonly referred to as MICE (Money, Ideology, Coercion, Ego). The agent of influence's task is to covertly support narratives, opinions, and actions that are beneficial to the country they work for, for which they establish contacts in politics, business, military, science, and media. Agents of influence operating in the public space aim to create a debate on specific political, social, or security and defense issues in such a way as to trigger the desired reaction (e.g., change of opinion or attitude) or action (e.g., protests, riots, or revolutions) in line with the client's expectations.

The task of the influence agent who operates within the structures of the state administration is to influence analytical and decision-making processes in such a way as to make it impossible to properly assess the situation and make appropriate decisions. This activity can also take the form of inspiration, in which the disinformed subject is directed to make a specific decision. This goal is achieved, for example, by creating an image of reality that aligns with the interests of the disinformer or by diverting attention from essential matters and replacing them with secondary ones (Świerczek, 2020).

Influence agentry has been one of the main methods of operation for Soviet/Russian intelligence since the Cold War, and there have been several high-profile cases of unmasked agents of influence from Soviet secret services during that time. These include Alger Hiss, an American official in the Department of State and the United Nations; Pierre-Charles Pathé, the founder of the "Synthesis" newsletter read by the French journalistic and political elite, who worked for the KGB from 1959; Arne Treholt, who joined as a young journalist and activist of the Norwegian Labor Party in 1967 then served as an advisor to the Minister of Trade and an official in the Ministry of Foreign Affairs; and Hirohide Ishida, a Japanese politician who was a close associate of two prime ministers and served as the minister of labor and transport.

In the 1990s, Richard Gott, a journalist for the British newspaper *The Guardian,* known for his radical views and sympathies for Ernesto "Che" Guevara, was accused of working for Russian intelligence as an agent of influence (Williams, 1994). He denied having connections to the KGB, HOWEVER, and was not convicted due to a lack of evidence. It is worth noting that the British were not only targeted by Soviet/Russian influence operations but also used similar methods in their fight against the Irish Republican Army (IRA). British intelligence believed that by supporting the leaders of organizations that preferred negotiations over armed struggle and cooperating with them, they could stabilize the situation in Northern Ireland at a lower cost (Edwards, 2021).

The relationship between an agent of influence and a "case officer" may be informal and not raise suspicion. In practice, the person used as a tool in the influence operation may not even receive direct instructions but only be subtly guided or inspired. This is particularly true for those who are motivated not by material gain or fear but by a sense of mission or ideology. This method is often used with journalists who, during meetings with intelligence officers disguised as "state officials", are intentionally provided with information, such as that pertaining to a particular country's internal situation.

There is a mechanism of triple addiction at work here. The first addiction is to inspire the agent of influence in everyday situations. The second addiction is if the recipient of the information finds it appealing, which increases the chances of publishing it in a newspaper or TV station. Finally, the third addiction is gratitude. The recipient of the information, in return for being provided with it, may offer repayment by disseminating it. As a result, the recipient becomes a proxy in a disinformation operation, inspired but not openly ordered by foreign intelligence.

The literature on this subject outlines three types of influence agents:

- Trusted contact. This is a person who holds a high position in a state institution, expert center, or media and maintains close relations with representatives of another state's structures, but only cooperates with them to a limited extent.
- Controlled agent of influence. This refers to a person who has been recruited by a foreign intelligence agency to carry out specific tasks in exchange for financial gain. These controlled agents of influence are typically identified

at a relatively young age, such as during their studies, and are provided with the necessary training before being deployed for operational purposes. In order to increase their chances of a successful career in state structures and avoid detection by local counterintelligence agencies, they may remain dormant for many years. These "sleeping agents" are activated when they reach a suitable status.

- Special contact. This refers to situations where the recruitment of a "controlled agent" is limited due to political reasons, such as allied relations between states. In this case, a special contact is used instead of an official agent of a foreign intelligence agency. The special contact does not work for the foreign intelligence agency but rather performs favors under the guise of "common interests of both countries" (*Active Measures*, 1986, pp. 81–83).

The Polish legal system does not have a specific categorization for "agent of influence" activities. However, such activities may fall under the category of espionage according to Article 130 of the Polish Penal Code. This article stipulates that "anyone who participates in the activities of foreign intelligence against the Republic of Poland shall be subject to imprisonment for one to ten years". In practice, however, obtaining a conviction under Article 130 for influence operation activities is more challenging than for classic espionage, such as in the case of stealing state secrets. This is because it requires demonstrating that a person was in regular contact with a foreign intelligence officer and accepted and carried out tasks on their behalf. This is relatively difficult to prove in a democratic state with the rule of law and freedom of speech.

The case of Mateusz Piskorski, the leader of the pro-Russian political party *Zmiana* (Change), provides an example of the procedural difficulties related to the classification of influence operation activities under Article 130 of the Polish Penal Code. Piskorski has been commenting on political events for Russian propaganda media such as *RT* and *Sputnik* for years. Since 2015, he has been the chairman of an openly pro-Russian group that describes itself as the "first non-American political party" in Poland. Members of *Zmiana* work with the Eurasian Movement, founded by one of Russia's main theorists and practitioners of information warfare, Alexander Dugin. Piskorski also created a "think-tank", the European Center for Geopolitical Analysis (ECAG), to promote a clearly anti-Western, anti-Ukrainian, and pro-Russian rhetoric that complies with the Kremlin's narrative, goals, and interests. In 2011, Piskorski participated in a propaganda conference organized by Libyan dictator Muammar Gaddafi. Two years later, he was invited by Bashar al-Assad to Syria and afterwards, he published a series of articles criticizing U.S. policy in the Middle East and North Africa. His organization, ECAG, also recruited "election observers" in unrecognized quasi-states that would not exist without Russian military interference. These include Transnistria in Moldova, Abkhazia and South Ossetia in Georgia, and Donetsk and Luhansk in Ukraine. Piskorski himself led an "observation mission" for the illegal referendum in Crimea in 2014.

Two years later, he was arrested and accused of working for Russian and Chinese intelligence. Despite compelling evidence of his role in Russian disinformation and propaganda activities, he was not convicted. After three years in custody, he was released and continues to engage in public activity, acting as a commentator and "Polish expert" for Russian propaganda media outlets (Wenerski & Kacewicz, 2017, pp. 27–30).

Piskorski is the most well-known example of an individual involved in disinformation activities in Poland inspired by the Kremlin, but he is not the only one. In May 2021, the Internal Security Agency detained Janusz Niedźwiedzki, who had ties to the leader of the *Zmiana* group. He attempted to establish contacts with Polish and foreign politicians on behalf of Russian intelligence. According to Polish law enforcement authorities, "his activities were part of Russian propaganda and disinformation projects aimed at weakening Poland's position in the EU and in the international arena" (*Janusz N*, 2021). Like Piskorski, he participated in influence operations by commenting on political events for Russian propaganda media outlets, spreading the Kremlin's narratives in Poland, and participating in "election observation" missions in Ukraine and Russia. His role was to undermine the results of the Dnipropetrovsk elections, which ended with the defeat of the pro-Russian candidate, and to legitimize undemocratic elections in Russia. His activities were financed by the Russian Peace Foundation, headed by Leonid Slutsky, Chairman of the Russian State Duma's International Affairs Committee. Niedźwiedzki also has connections with pro-Russian organizations in Poland and Europe, including the Night Wolves motorcycle gang, which is engaged in Russian influence operations (Shekhovtsov, 2021).

A similar role to that of agents of influence is played by "useful idiots", that is individuals who spread disinformation and propaganda messages thoughtlessly or unconsciously. Their activities are not usually the result of being tasked or inspired by a foreign intelligence officer but rather stem from their personal views, knowledge (or lack thereof), sympathy, beliefs, and ideology. Another category of disseminators of disinformation are trolls. These are individuals who are commissioned and paid for their activities, which a focus on posting and commenting on social media in line with the "needs of the disinformer".

The most well-known Russian "troll factory" is the Internet Research Agency (IRA), which has operated from Saint Petersburg since 2013. Its owner, Yevgeny Prigozhin, is a close associate of Vladimir Putin. The IRA's monthly budget exceeds 1 million euros, which allows it to employ around 1,000 people who spread Russian narratives and false information, reinforcing extreme social and political attitudes abroad (Legucka, 2019). According to a former IRA employee, the troll farm is divided into departments responsible for specific social media platforms (such as Facebook, Twitter, or YouTube) and specialized forms of disinformation, such as creating memes or collecting compromising materials (i.e., *kompromats*). According to the British organization of investigative journalism Bellingcat, the Russian Ministry of Defense and the military intelligence (GRU) are responsible for the operations conducted by the IRA (*Putin Chef's*, 2020).

Bots perform the same task as trolls in disseminating disinformation but are accounts created and managed by computer algorithms. They are widely used on social media to automate the spreading of disinformation. According to research conducted by the NATO Center of Excellence for Strategic Communications, following NATO's decision to deploy multinational allied battalions to Poland and the Baltic States, 84% of tweets in Russian and 46% in English that negatively referred to the presence of NATO troops on the Eastern flank were produced by bots (Fredheim, 2017–2021). Despite a downward trend since 2017, largely due to actions taken by social media owners, bot activity remains high. This demonstrates the increasing use of modern technologies for automated social engineering as part of information warfare in cyberspace. Social media platforms still struggle to detect and remove false content produced by bots, particularly in the case of non-English languages. An experiment conducted by NATO StratCom CoE showed that the effectiveness of individual websites in detecting and combating inauthentic accounts was as follows: 35% for Twitter, 21% for Facebook, 14% for Instagram, and 11% for YouTube (Bay & Fredheim, 2019, p. 22).

The credibility of the source providing manipulated or false information is critical to the success of an intelligence disinformation operation. Therefore, at the initial stage, the source provides only true information to build its reputation. Once it has gained the trust of recipients, it changes its approach by mixing true information with manipulated or completely fabricated information. In such operations, the following methods are used:

- Distraction, or the provision of specially prepared information to redirect the subject's interest to other areas.
- Misleading, or the creation of a false image of reality to influence the perception of a disinformed entity.
- Convincing, which involves efforts aimed at the authentication of disinformation by raising its credibility among public opinion.
- Disinformation by suggestion, or the indirect shaping of an image of a phenomenon in a way that is favorable to the disinformer (Rusbridger, 1993, p. 66).

The distinguishing feature of disinformation carried out by intelligence services is its secret nature. This applies to influence operations directed at state institutions as well as those targeting entire societies. Specialized methods, techniques, and tools are used to mislead the recipient(s) and make messages more credible. To mask the involvement of the inspirer of the disinformation, a system of intermediaries is put in place. In the case of Russia, this system can be called a "matryoshka system" (Świerczek 2018, pp. 210–228), which allows the actual sender of the manipulated message to be hidden many times over. The effect of this type of action is to evoke a specific reaction from the recipient in line with the intentions and interests of the disinformer. The true art of it is to ensure that the object of manipulation remains unaware of it.

3 Military Disinformation

The use of disinformation as a tool in politics, diplomacy, trade, or warfare is an age-old strategy. The advantage of information is an indispensable component of success in times of war and peace, as well as in "gray zone" or "hybrid" conflicts. The essence of these conflicts is the use of a combination of military and non-military means to keep the confrontation under the threshold of war (Hoffman, 2009; Piotrowski, 2015, pp. 7–38). In such a context, disinformation is an element of information warfare (infowarfare), which is defined as "actions aimed at protecting, using, damaging, or destroying information or its resources, as well as contradicting information in order to achieve significant benefits, goals or victory over an opponent" (Schwartau, 1996). It encompasses offensive and defensive actions necessary to gain an information advantage over the enemy and to achieve the intended military and political goals. Military deception (MILDEC) is one of the tools of infowarfare and is understood as the deliberate transfer of specially prepared (manipulated or fake) information (e.g., via documents or demonstrations of military actions) aimed at misleading the opponent regarding real intentions, plans, and undertakings to achieve military advantage. Its application is crucial for obtaining the element of surprise, securing actions, and minimizing the losses of the used forces and resources.

Military disinformation is a tool of strategy that has been known to strategic theorists and practitioners since ancient times. Its operational use, however, was significantly developed during the Second World War. For instance, in 1942 the Allies conducted "Operation Torch", which suggested that they were preparing a landing in Norway or France. The aim was to deceive the French collaborationist government and prevent the strengthening of German forces in North Africa, while the real target was Algeria and Morocco. Another well-known example is "Operation Mincemeat" (1943), a two-tier disinformation operation conducted by British intelligence. The operation name suggested that the planned Allied landing on Sicily ("Operation Husky") was a sham and that the real targets were Sardinia and Greece ("Operation Brimstone"), which led to a change in German and Italian defense plans. The actual landing target was Sicily, and the strikes against Greece and Sardinia were simulated. In 1944 operations "Bodyguard" and "Fortitude" suggested preparations for Allied forces landing in the Pas de Calais area and off the coast of Belgium, while the real target of "Operation Overlord" was Normandy. The purpose of this disinformation was to force Germany to disperse its activities held in various theaters (Hughes-Wilson, 2002).

Military disinformation aims to cause chaos in the enemy's command and control system (C2) by providing false information, leading to the adversary's incorrect assessment of the situation. It has a specific goal and implementation plan, such as prompting the opponent to change their defense concepts, shifting the location of military groups, or redirecting the enemy's forces and resources to a mock strike area. The success of disinformation activities is determined by the opponent's actions, which align with the deliberately created picture of the operational situation. The following methods are used in the implementation of MILDEC operations:

- Disinformation by intelligence assets. This involves providing fabricated information to the enemy (such as documents, operational plans, orders, reports, decisions, diagrams, and maps) through the use of human intelligence assets (such as agents, double agents, and offerors).
- Disinformation through inspiration from the environment. This is the process of disseminating false information by spreading rumors among the enemy troops, armed groups like rebels, resistance forces, guerilla fighters, and the local population.
- Disinformation via the mass media. This involves spreading false information to the press, radio or television or on social media.
- Radio-electronic disinformation. This refers to the transmission of false commands, orders, and reports through information and communication technologies (ICT), as well as the use of fake radio-electronic and digital communications to simulate activities (such as preparation to start operations in a different direction). It also involves emitting misleading electronic, audible, and visual signals to authenticate disinformation.
- Operational masking. This tactic is aimed at hindering the enemy's ability to make informed decisions and conduct effective military operations. It involves concealing troops and defense infrastructure from enemy reconnaissance capabilities and using dummies and mock-ups.

Military disinformation plays a supporting role and is one element of military information operations (INFOOPS). It is also a component of strategic communication (StratCom), which is defined in NATO as the "integration of communication capabilities and infoops with other military activities, in order to understand and shape the information environment" (MC 0422/6, p. 6). These capabilities include public diplomacy and military public affairs. INFOOPS also include measures such as military reconnaissance, operational security (OPSEC), electronic warfare (EW), psychological operations (PSYOPS), and the physical destruction of the opponent's information systems.

According to NATO standards, tools for strategic communication and information operations involve:

- Strategic communication. This is the coordinated and tailored communication activities and capabilities that serve two functions. Firstly, it integrates coalition information efforts to secure vital interests and goals and promote coalition cohesion. Secondly, it provides advice and coordination for undertakings that affect information and information systems, including the behavior and capabilities of these systems, to achieve the desired effect.
- Public diplomacy. This involves activities in the field of civil communication, which is supplemented by activities and supporting tools that promote awareness and build understanding and support for NATO policies and operations.
- Public affairs. This involves the timely, accurate, active, and reactive involvement of NATO's civilian sector in reporting through the media about Alliance policies and ensuing activities and operations.
- Military public affairs. This aims to promote the impact of NATO's military objectives among the facilities to raise awareness and promote a better understanding of the military aspects of the Alliance's functioning. It includes the planning and implementation of appropriate relations with the media, internal communication, and relations with society.
- Information operations. These ventures are coordinated by a staff cell and rely on analyzing the information environment, planning, integrating, and evaluating information activities to achieve the desired effects of influencing the will to act. They also rely on an understanding of the situation and the capabilities of the opponent and other approved objects of influence. They support the achievement of the operation's objectives and strategic communication.
- Psychological operations. These involve a scheduled process of transmitting prepared content using various methods and means of communication to selected targets to bring about the desired change in perception, attitudes, and behavior, which will facilitate the achievement of intended political

and military objectives. The PsyOps units' tasks include: (1) weakening the enemy's morale and its abilities, including combat abilities; (2) consolidation of friendly/neutral attitudes of groups and communities not directly involved in the conflict; (3) establishing cooperation with indecisive or neutral circles and gaining their support; (4) participating in undertakings related to operational masking and operational security (OPSEC); (5) collecting, analyzing, processing and disseminating data about the enemy and the area of operations within the integrated reconnaissance system; (6) supporting undertakings carried out by cells of the strategic communication system preventing the opponent's psychological operations; and (7) developing and improving one's own and allied database.

– Military reconnaissance. This involves gathering, processing, and disseminating information about the enemy.
– Operational security. This refers to ensuring sufficient protection for military operations or activities through passive or active means to prevent the enemy from accessing critical information on the deployment, capabilities, and intentions of one's own troops. For this purpose, disinformation is also used to mislead adversaries and conceal the actual intentions and actions of one's own troops.
– Electronic warfare. This involves military operations that aim to identify and disrupt enemy electronic systems and emissions, while also ensuring optimal conditions for the use of electronic systems by friendly troops. Electronic warfare is a crucial component of information activities, involving reconnaissance, electronic countermeasures (interference), and targeting the enemy's command, communication, IT, and reconnaissance systems.
– Physical destruction of information systems. These are strikes performed to deprive the combat capability of the key elements of the enemy's command, control, communication, computers, intelligence, surveillance, and reconnaissance systems (C4ISR).

The above definitions primarily represent the Western perspective on the elements of information warfare. They may not align with how authoritarian regimes interpret and apply them. Russian theorists of information warfare distinguish its two components: (1) information-technical and (2) information-psychological. The former involves integrated actions against the adversary's entire ICT infrastructure, including communication channels, radio-electronic means, and command and control systems of their armed forces (Thomas 2010; Giles 2016). The latter refers to a set of activities aimed at gaining and maintaining an information advantage over the enemy during military operations.

The purpose of Russian information-psychological operations is to disrupt the enemy's information resilience by maintaining constant psychological pressure on the adversary (Nowacki, 2004, pp. 144–147). Russian theorists do not distinguish between military and non-military, technological (cyberspace) and social (information space) order, or times of peace and war (Darczewska & Żochowski, 2017). Unlike the Western approach, Russians do not consider cyberspace as a separate strategic theater of military operations alongside air, sea, land, and space. Instead of using the term "cyberspace", they use the term "information space". Russia's cyber-capabilities are just another tool of information warfare alongside intelligence, counterintelligence, disinformation, propaganda, electronic warfare, disruption of communication and navigation, psychological pressure, and destruction of enemy ICT resources.

Military disinformation can be either "passive" or "active". Passive disinformation aims to hide the adversary's real intentions and abilities, while active disinformation provides fabricated "evidence" of alleged intentions and abilities. Disinformation can also be based on ambiguity (type "A" deception) or on misleading (type "M" deception) (Caddell 2004, pp. 6–7). In the case of type "A" deception, the aim is to cause general confusion in the enemy's headquarters through simulated military or diplomatic activity, increased radio communication, or simulated negotiations in order to gain time to prepare an armed operation. An example is Japan's diplomatic activity during the preparations for the attack on Pearl Harbor in 1941, which made it difficult for American military analysts to assess the situation and Tokyo's real intentions (Wohlstetter, 1962). In the case of type "M" deception, the goal is to mislead the opponent about the planned course of action and to reassure them that they are correctly assessing the situation. An example is "Operation Bodyguard", which masked the preparations for the Normandy landings in 1944 (Howard, 1990).

Disinformation can also be classified as "offensive" or "defensive". Offensive disinformation aims to mislead the opponent about future intentions and force them to fight under unfavorable conditions. Its goal is to achieve surprise and gain initiative. Defensive disinformation activities, on the other hand, are aimed at opponents who have the advantage (initiative) and aim to divert their attention and efforts from the actual goals and plans of the operation. The main goal is to improve the security of one's own activities and create conditions for the successful completion of assigned tasks (Wrzosek, 2005, pp. 75–76). Disinformation can also be carried out at different levels of command within an operation. In this order, its division is as follows:

- Strategic disinformation. This is carried out to mislead the enemy regarding the time, place, strength, and intention of a campaign or operation.

It encompasses matters related to the training system, organizational structure, deployment, level of combat readiness, methods and timing of mobilization, operational and strategic development of the armed forces, composition, equipment, and combat readiness of deployed troops during peacetime, as well as the command and control system.

- Operational disinformation. This involves activities and measures to mislead the enemy regarding the conduct of operations. They should be consistent with, and a logical consequence of, false information communicated at the strategic level. The main purpose is to conceal the preparations and intentions to conduct the operation. This applies particularly to the readiness of separate components of the armed forces planned for use, areas of operational and strategic reserve dislocation or mobilization, as well as the readiness of precision destruction means, areas for the development of command posts, and the operating mode of ICT systems. It also includes directions of military maneuvers and the movement of logistic support shipments.
- Tactical disinformation. This refers to all activities and measures used to mislead the enemy in the area of operations. Actions at this level include protecting critical information, masking, concealing, and simulating (Modrzejewski, 2015, pp. 92–93).

Military disinformation is a specialized form of deception primarily directed at the enemy's armed forces, especially its command, communication, information, computer, and reconnaissance systems. Its aim is to create a credible but false picture of the strategic, operational, and tactical situation. Military disinformation employs methods typical of intelligence activities, such as the use of agents and mass media. It is also understood as one of the forms of INFOOPS or PSYOPS and considered an element of a broader set of strategic communication tools aimed at the enemy's armed forces, decision-making centers, and society. The success of military operations often depends on the attitude of the population living in the war zone, making the informational "struggle for hearts and minds" crucial in the era of asymmetric, unconventional, and hybrid conflicts (Lennon et al., 2003; Stubbs, 2004). In the 21st century, information should therefore be regarded as a weapon.

CHAPTER 6

Identifying Disinformation That Targets Mass Audiences

Disinformation can be a component of operations conducted by intelligence services or specialized military units. Their activities may be directed at specific political or military decision-making centers or entire societies. Regardless of the object of disinformation, its goal is always to trigger changes in the consciousness of recipients and disturb their cognitive abilities. This can then lead to a change in attitudes and cause a specific social, economic, or political reaction (Aronhime et al., 2021). Disinformation activities directed at a mass audience may aim to:

– Sow doubts and manipulate public opinion.
– Influence social and political attitudes.
– Distract public debate.
– Polarize the political situation.
– Undermine the value system of a state or community.
– Weaken the cohesion of a state or group of states.
– Undermine trust in public institutions and the media.
– Spread ideologies and discredit the opposing ideologies.
– Inspire chaos, division, and conflicts.
– Destabilize a society, state, or community.
– Undermine the integrity of the government, constitutional principles, or political (decision-making) processes.

Despite the previously detailed characteristics that differentiate misinformation, disinformation, and malign information, these three classifications still have many things in common. All three are tools in the hands of a manipulator, and they can be used to elicit a certain response by a recipient while camouflaging the manipulator's real intentions.

Counteracting and combating disinformation are difficult in open societies where fundamental freedoms like freedom of speech are valued and protected. In such circumstances, it is difficult to punish or publicly stigmatize those responsible for creating, reproducing, and disseminating manipulated or false information. This is due to several reasons.

First, in some cases, disinformation can be created as a result of a person's insufficient intellectual or social competence. It can therefore be reproduced through inadvertent action and without malevolence. In democratic societies, presumption of innocence is a fundamental component of the rule of law and freedom of expression. It has also been the foundation of efforts by the democratic community to de-penalize journalistic offenses (i.e., derivatives of misuse of information) and strongly present in the activities of the Organization for Security and Co-operation in Europe (OSCE).

Second, disinformation often has a basis in fact but relies on the author's interpretation, opinions, or comments to create a message that is inconsistent with reality. A dilemma therefore arises between the potential harm that false information may inflict on an individual or society and the right to freedom of expression. Both values are the subject of legal protection in modern democratic societies; but while it is often difficult to prove that harm was derived from such actions, the right to free expression remains highly valued and protected within the free world.

Third, disinformers who are caught spreading false or manipulated information often invoke their right to freedom of speech, which is guaranteed in Article 11 of the Charter of Fundamental Rights of the European Union. Despite the fact that EU institutions recognize that disinformation not only "undermines trust in institutions, traditional and digital media, and damages democracies by preventing citizens from making informed decisions" but also "impairs freedom of expression" (*Tackling online* 2018, p. 1), they have not put forth proposals for administrative and regulatory measures or an in-depth reform of applicable laws to address this threat. While the integrity of public debate is safeguarded by established institutions and the strength of civil society, this task is more difficult in the case of online platforms, which are seen as a "space for unlimited freedom of expression". To effectively counteract the phenomenon of disinformation, individual countries have independently introduced legal provisions aimed at minimizing its effect.

In 2017, the German parliament adopted the Network Enforcement Act (NetzDG) to improve law enforcement in matters relating to social media. According to the legislation, digital platforms must provide users with the ability to submit complaints and they must remove content that is illegal (within 24 hours) or contains false messages (within 7 days). Yet this does not address the fact that in the digital age disinformation can be reproduced countless times within 24 hours and its effects may not be reversed. Apart from the obvious weakness of such late intervention, some German political parties consider this law to be contrary to freedom of expression and the constitution.

In France, a similar law requires internet service providers to disclose information regarding entities funding the promotion of political content during an election period. It also prohibits the dissemination of inaccurate, misleading, or suspicious statements that could impact the fairness of elections, with the exception of satirical content and programs. Moreover, the French regulatory authority has the power to reject a media license application if it determines that the organization in question poses a significant threat to values such as human dignity or freedom of thought and opinion, or if it jeopardizes essential national interests, including the proper functioning of public institutions. The act also imposes obligations on internet platform operators toward users and state authorities, such as:

- Providing tools for reporting false information.
- Ensuring the transparency of algorithms used.
- Promoting content from news agencies and audiovisual communication services.
- Disabling accounts promoting false information on a massive scale.
- Informing users about sponsored content of general interest and disclosing the amount of money received to promote it.
- Providing information on activities in the field of media and information education (Ogrodowczyk et al., 2020, pp. 24–28).

Disinformation targeting mass audiences refers to large-scale information warfare activities that aim to achieve a strategic goal through coordinated disinformation operations. It can also be defined as "a series of coordinated information operations conducted by a foreign entity aimed at influencing a specific group of recipients with the intention of achieving specific goals and benefits by the initiator" (Twetman et al., 2019, p. 5). A crucial aspect of this phenomenon is the unlawful use of information against society, the principles of democracy, and open public dialogue. Disinformation campaigns exploit weaknesses in media systems, cognitive biases, and public opinion-shaping processes. They are often conducted covertly, with the help of local proxies to make it challenging to identify the real, often foreign, perpetrator (Maurer, 2018, pp. 171–188). The objectives of such campaigns include eroding trust in public institutions, promoting social polarization, radicalizing public debate, and excluding certain social groups.

Disinformation campaigns often revolve around a "topic" that is easy to remember, evokes associations with the targeted entity, has significant social or political importance, and elicits strong emotions. They may also be based on issues used in several operations at the same time. Such campaigns often center around current political events (such as elections, legislative activities, public scandals, or other contentious issues) that provoke public objection. The choice of the subject of disinformation campaigns is not random but rather the result of operational diagnosis, which is akin to "market research" aimed at determining: (1) What topics are important to a given group of recipients? (2) Which of these topics evoke the greatest emotions? (3) How will the manipulation of messages on specific topics affect the political and social situation in the target country?

Regardless of the subject of the disinformation campaign and its specific objectives, the overarching goal remains to instill doubt in the recipient, create a power-society discrepancy, and promote an intentionally crafted alternative way of interpreting reality. Due to the limited legal and physical means available for combatting disinformation within a democratic system of values, building social resilience is the primary defense mechanism. Its foundation is the individual skills of each recipient of information in recognizing falsehoods and manipulation, properly selecting reliable and credible sources, and debunking lies online.

1 Recognizing Disinformation

There are currently approximately 4.57 billion active internet users worldwide, each producing an average of 1.7 megabytes of data per second. This amounts to a staggering 2.5 trillion megabytes of data generated daily. With socio-economic dynamics and technological acceleration expected to continue, these numbers are only set to increase, driven by faster computing powers, new applications like the Internet of Things, and the capabilities of quasi-human or superhuman machines fueled by artificial intelligence (Kupiecki, 2020a, pp. 472–497). These developments will create new opportunities for variously motivated entities operating in the domain of disinformation, who will attempt to create alternative realities to garner political or business benefits or harm individuals, nations, and the international community.

In the last two decades, as much as 90% of the world's data has been produced. This number is projected to increase to 463 exabytes by 2025 (Bulao, 2020). With so much information available, how can individuals and nations navigate it and evaluate its credibility? How can they distinguish real information from manipulated or false information? To answer these questions, it is essential to study the reliability of information sources and producers, particularly in a world where the monetized "click-through rate" is prioritized over reliable and credible information. The business model of news media outlets, which relies on advertising profits, often leads to the rapid publication of new content at the expense of quality and accuracy. This rush to publish can limit the time available to verify information, while sensationalized, emotionally manipulative headlines (clickbait) capture readers' attention and are used to generate ad revenue (Głowacka et al., 2019, p. 4).

In the modern world, true information (facts) coexists with falsehoods, and this makes it increasingly difficult for recipients to discern which is which. The ability to freely choose what to believe is a great privilege of freedom, but it requires basic knowledge and intellectual skills. Computer algorithms today can collect data on users' online activity, analyze their preferences and interests, and then send suggestions that align with their anticipated expectations or "needs". The current technological possibilities in the field of profiling offer a wide range of opportunities for social engineering, including direct and effective influence on the attitudes and political decisions of individuals and states. These practices raise serious ethical questions, as illustrated by the role of Cambridge Analytica (CA) in the 2016 U.S. presidential election, the Brexit referendum, and the secession referendum of Catalonia in 2017. CA obtained data from 50 million Facebook users without their awareness, which allowed the company to segment and target voters. Based on this, CA developed specific content and methods of distribution in social media to reach voters and influence their attitudes (Boldyreva et al., 2018, pp. 91–102).

In the 21st century, cyberspace has become a modern field for civil and military interactions, including information warfare, cyber-attacks, data and intellectual property theft, false identity usage, cyberespionage, tracking, and surveillance. New technologies offer wide possibilities for creating an alternative reality by means of mass production and distribution of manipulated or completely false information. Authoritarian states use these techniques systematically and over long periods to influence social attitudes and decisions made by other participants in international relations. These actions can also be used to provoke riots, social unrest, and even armed conflicts. It is therefore important to consider:

- How can we mitigate the disinformation threat?
- How do we distinguish real information from fake news?
- How can we assess the credibility of information sources and make appropriate selections?
- How can we immunize ourselves to our surroundings against massive disinformation attacks?
- How do we educate adults and youth on these topics from the earliest stages of education?

The group of targets for disinformation includes politicians, diplomats, soldiers, business people, experts, analysts, journalists, commentators, academics, and other social groups. Each of these groups must possess the necessary skills and tools to defend themselves against false information. The success of disinformation is measured by the degree to which the manipulated message is recognized as 'true' by its recipients. This not only alters their perception of the phenomenon in question but also perpetuates the disinformation in a wider group of recipients. Anyone can therefore help mitigate the harmful impact of disinformation by checking the credibility of information and its source before sharing it, or by exposing information falsehoods if they have already been disseminated. Manipulated individuals – often enjoying social authority – who spread false information not only expand the scope of the destructive influence of disinformation but also legitimize its "truthfulness" among their followers. According to research conducted by the MIT Media Lab, real information takes about six times longer than fake information to reach 1,500 people on Twitter. Untrue news has nearly a 70% greater chance of being shared than news based on facts, particularly when it comes to politics (Metz, 2008). To counteract the phenomenon of mindless forwarding of unverified information, Twitter has introduced a mechanism that warns users and suggests reading linked articles before sharing. The "trap of authority" and the "trap of the attractiveness of fake news" additionally reinforce the functioning of recipients in "filter bubbles", which display personalized messages that may not always reflect reality.

Another concerning trend in information operations is the questioning of the credibility of scientific research and expert opinions by self-proclaimed pseudo-authorities. Bloggers, vloggers, and celebrities often lack relevant knowledge, skills, and qualifications yet they are very popular on social media platforms. With thousands of followers on Facebook, Instagram, YouTube, or Twitter, they can influence society to a greater extent than reliable and credible individuals and institutions. Targeting disinformation activities at such groups will be most effective in operations aimed at destroying the foundations of mutual social trust.

2 The RESIST Model: Recognition and Analysis of Disinformation

Recognizing and analyzing disinformation requires knowledge of the principles, methods, and techniques used to mislead the recipient. The British government communication service has developed a useful model called RESIST, which includes tools for identifying and analyzing disinformation, providing early warning, conducting proactive and reactive strategic communication, and verifying its effectiveness. This model was primarily created for public institutions and the private sector. It is one of many proposals on how to combat disinformation, hate speech, and harmful marketing campaigns (known as "black PR") by organizing special teams or departments responsible for monitoring, analyzing, and responding to such phenomena. The British government's initiative aims to standardize the methods and means of combating disinformation and facilitate cooperation in situations of crisis or external aggression that employs hybrid techniques. The RESIST model can also serve as a guide to designing an organization's information or communication policy. Its flexible solutions allow for adapting to the individual needs and specificities of a particular entity.

The British model of recognition, analysis and response to disinformation (RESIST) contains the following elements:

a) Recognize. This provides an overview of the current information environment, specifically the vast amount of information available online and the difficulties it poses for individuals trying to navigate it effectively. It helps to explain the distinctions between misinformation, disinformation, and harmful information, as well as the potential negative effects they can have on those who consume them. As part of the diagnosis, a checklist is created that is used to determine the likelihood of the manipulation or falsification of information. It involves the following control questions:
   - What are the goals of disinformation?
   - What are the techniques of disinformation?
   - How are disinformation techniques combined to achieve an effect?

b) Early warning. This begins with a review of available tools that can be used to monitor the information space. The assessment enables organizations, groups, or individuals to prioritize their actions and identify areas vulnerable to disinformation. By identifying targets, audiences, disinformers, and risks, it helps to focus on monitoring key weaknesses and take necessary measures to mitigate the impact of disinformation. The relevant control questions are:
   - How can I prioritize digital monitoring?
   - How can I build an individual set of tools for digital monitoring?
   - How can I use digital monitoring to assess the facility's susceptibility to potential threats?

c) Situational insight. This explains how disinformers shape the information environment and emphasizes the importance of situational context analysis. The analysis can be conducted systematically through regular reports (daily, weekly, or monthly) or in response to emerging threats and issues. It focuses on the following control question:
   - What is the situational context of disinformation activities and how can it be used to support a rapid response?

d) Impact analysis. By analyzing the methods and techniques used in disinformation campaigns, it is possible to not only understand the meaning of specific disinformation operations but also predict likely future campaigns and their impact on recipients. The control questions are:

- What is the likely purpose of the disinformation?
- What is the likely impact of the disinformation?
- What is the likely extent of the disinformation?
- How should disinformation be prioritized?

e) Strategic communication. This contains a set of key methods and tools for proactive, active, and reactive communication. Public and private sector entities should consider these tools when developing and implementing their communication policies and strategies. It is necessary to identify effective channels to reach target groups and use "friendly voices" to increase the credibility and reach of the entity's communication activities based on various situational scenarios. The control questions are:
   - What should the public response to disinformation look like?
   - What is the approval process like?
   - What are the available response options?

f) Track outcomes. This step enables users to evaluate the effectiveness of their own strategic communication. It starts with the following control questions:
   - How should information on the disinformation campaign be recorded and shared ?
   - How can I evaluate my own actions and understand the conclusions drawn? (*RESIST Disinformation,* 2020, p. 4; *RESIST 2*, 2021, pp. 6–7)

In a simplified version, the RESIST model can also be used by individuals or analytical networks whose abilities to identify, analyze, and debunk false information contribute to the system of social resistance to disinformation. The knowledge of basic methods, tactics, techniques, and mechanisms of disinformation can improve their capabilities. Such knowledge makes recipients better prepared to recognize manipulations of narratives, facts, or contexts. The basic components of disinformation can be summarized with the acronym FIRST:

- Fabrication. This is the manipulation of a message's content through the use of false text, documents, or pictures.
- Identity. This concerns concealing or stealing a person's identity to use it on fake social media accounts.
- Rhetoric. This is the use of argumentation based on false information or offensive content in messages, for example, by trolls commenting on social media posts.
- Symbolism. This is the malign use of symbols to strengthen a communication message by comparing a politician to a controversial historical figure, for instance.
- Technology. This involves the use of technological advantages, like bots, which can automatically produce false messages on a mass scale (*RESIST Disinformation*, 2020, p. 9).

Although the above elements provide a general understanding of how false information is created and disseminated, their purpose is primarily to raise awareness among recipients about the need to verify information consciously. Unlike professional, reliable, and credible press offices, information published on the internet is often not subject to a comprehensive verification and approval process. In principle, each user can create their own "media" (such as a website, blog, vlog, or portal) and publish any content they want. The quality and credibility of such content must be evaluated independently by the recipient, who may not always have sufficient knowledge about the subject matter. Technical capabilities to manipulate content, modify photos, impersonate others, commit identity theft, or create fake accounts are widely available. However, this does not mean that all users employ them with bad intentions. Maintaining a well-understood skepticism and criticism toward unknown sources of information can reduce susceptibility to disinformation. A "classical methodology" for conscious verification of information was characterized by the French researcher specialized in this phenomenon, Vladimir Volkoff, who distinguished the following information manipulation methods:

- Negation or inversion of facts.
- Combining truth and lies. This is used in a situation where the basic circumstances are generally known but the details have not been made public.
- Modification of a motive. This refers to a tactic used by disinformers where they change an element whose value was only known to them before a certain event. Once the situation is recognized, the disinformers change their motive of action to something that is socially acceptable and compliant with the norms of their environment.
- Modification of circumstances. This refers to a change in circumstances that are difficult to confirm unequivocally, particularly at an emotional level, such as judgments, feelings, and relationships. This method introduces chaos into the assessment of the situation, making it harder to determine the truth.
- Blurring. This has the same effect as the previous method but consists of flooding the main information with a large number of irrelevant facts.
- Camouflage. This is the opposite of blurring and involves breaking down the main information into such small details that important elements are lost.
- Interpretation. This is used when the facts are indisputable. They are then interpreted in such a way as to achieve the desired effect.
- Generalization. This tactic constructs a universal principle based on an individual example.
- Illustration. This is the use of an individual event as a "legitimate" illustration of a wider social phenomenon.

- Unequal representation. This is the manipulation of the quality and popularity of an information source, which can influence the difference in perception of information between an opinion-forming and popular source versus a little-known and niche source.
- Equal representation. This method is used in the final phase of a disinformation campaign, when the majority of the target group is already convinced of the theses promoted by the agents of influence. Its primary aim is to consolidate the already widely accepted opinion and close the topic (Volkoff, 1999a).

Despite the passage of time, the disinformation methods described by Volkoff continue to be used. Their universal nature is based on the cognitive limitations of human beings. Manipulating the content of a message falsifies the image of reality that reaches the recipient, influencing perception, shaping opinion, and determining actions. Technological capabilities for collecting big data about internet users facilitate the selection of target groups for disinformation campaigns. Combined with the ability to mass disseminate false content using computer algorithms, disinformation can be considered a unique "weapon of mass destruction of human minds in the 21st century". New technologies have expanded the possible methods, techniques, and mechanisms of disinformation, which include:

- Astroturfing. This involves presenting top-down agitation campaigns as civic initiatives or falsely attributing a given message to other entities to authenticate them.
- The bandwagon effect. This is a cognitive effect which reinforces a specific opinion or belief because it is shared by others. Social media users are more likely to share articles that have been shared by many others, regardless of their content.
- False connotation. This a situation in which the title, lead, photos, or graphics used do not correspond to the content of the message.
- False context. This is a situation in which content is based on facts but is placed in a manipulated information context.
- Filter bubble. Filtering algorithms use personalized access to information' based on a user's search history and social media activity. This puts the user in a situation where they are more likely to see content that corresponds to their previous online activity.
- Leaks. This is the deliberate distribution of information obtained illegally, which includes, for instance, the publishing of classified documents or the theft and publication of private or business correspondence written by persons holding state positions.

- Malign rhetoric. This is the use of slanderous and false accusations used to disrupt public debate.
- Manipulation. Modifying the content of information can be used to alter its meaning.
- Misappropriation. This involves falsely attributing an argument, statement, or position to someone.
- Satire and parody. This involves making fun of people (e.g., using memes) or perpetrating narratives or opinions with the aim of undermining their importance.
- Sock puppets. This is the creation of a fictional debate between two (or more) entities using new technologies. For example, this can be done by creating fake social media accounts and conducting discussions between them.
- Trolling. This involves the deliberate provocation and aggravation of discussions on social media by placing controversial, offensive, or emotional comments to provoke outrage in recipients and draw them into a discussion (*RESIST Disinformation,* 2020, pp. 21–22; Brodnig, 2017).

The scale of the problem around disinformation is best illustrated by the amount of false information circulated about the coronavirus pandemic. According to a study by Carnegie Mellon University (CMU), 82% of the 50 most popular Twitter accounts duplicating fake news about COVID-19 were bots (Huang & Carley, 2020). The consequence of this situation is the development of an "infodemia", as described in the first part of this book, resulting from the creation and reproduction of huge amounts of misinformation, disinformation, and conspiracy theories that questioned medical facts or assigned the "invention" of the virus to Western states or pharmaceutical companies for their political and commercial goals (Constantinou et al., 2021, p. 4).

In Poland, for example, approximately 50% of the population agrees with conspiracy theories about the coronavirus, with between 43% and 56% of respondents aged 18–74 believing in them, depending on the theory (Duplaga, 2020). Based on COVID-19 disinformation, numerous anti-vaccine movements emerged that questioned the lethality of the virus and the effectiveness of the vaccines. The lack of social immunity to such content translated into an insufficient level of vaccination of the population, leading to lower collective immunity to the virus. As a result, the number of victims of the pandemic increased to over 106,000 in Poland and over 5.7 million globally (as of February 2022). It is important to note that not every case of death or coronavirus infection was the result of omission or fear compounded by disinformation. It undoubtedly contributed to these statistics, however.

It may seem that individual users are helpless when faced with new technologies in the battle against disinformation. However, knowing the methods, techniques, and mechanisms of disinformation is very useful in identifying and analyzing materials with manipulated content. An example of this is a short study of source material containing a false historical narrative regarding the introduction of martial law in Poland in 1981. On December 13, 2020, the 39th anniversary of this tragic event, the Russian disinformation outlet *Sputnik Polska* published an article aimed at depreciating the importance of these events and trying to change the perception of Poles.

To do so, the author employed the following techniques:

- Satire and parody. Ironizing about the alleged "cruelty of the communist leader General Wojciech Jaruzelski", whose greatest crime in the author's opinion was "the lack of a morning TV program for children", the author attempted to diminish the significance of the general's actions. Additionally, the author mentioned her "gratitude to Jaruzelski" for not having to attend school for three weeks.
- False context. In the article, the author deprecated the brutality of the methods used by the communist repression apparatus, ignoring the facts of 40 deaths during martial law with 60 additional people wounded and over 10,000 interned.
- Providing data without a source. The author cites alleged public opinion polls suggesting that Poles have recently changed their attitudes toward martial law to be more positive, but no source is referenced. The hyperlink provided leads to a text about the conflict in Nagorno-Karabakh instead of a relevant source on the topic.
- Justification. The author justifies the decision to introduce martial law by suggesting that, otherwise, the Soviet army would have intervened in Poland according to the "Brezhnev Doctrine." While historical knowledge confirms the risk of such an intervention, it does not justify the brutality of the repressive apparatus of the Polish communists, who had begun preparations to pacify Solidarity well before December 13, 1981.

Russian disinformation operations often employ historical manipulations (Juurvee et al., 2020; Domańska & Rogoża, 2021; Legucka & Kupiecki, 2022), particularly regarding the Second World War, which is referred to as the "Great Patriotic War" in Russia. During the 80th anniversary celebrations, Russian authorities, including Vladimir Putin, suggested that Poland was responsible for provoking the Third Reich to attack, depreciating the significance of the secret Molotov-Ribbentrop Pact and pointing to the "Munich Agreement" of 1938 as the main cause of the conflict. By denying the Red Army's complicity in the aggression against Poland on September 17, 1939, the myth of the "liberator" and "defender of the Slavic nations" is perpetuated. Attempts to combat this false narrative are met with accusations of "Russophobia." By shaping an alternative interpretation of history, the Kremlin influences not only external recipients but also the historical awareness and sensitivity of Russians themselves.

According to research by the Center for Polish-Russian Dialogue and Understanding and the Levada Center, most Russians view the Red Army's intervention in Poland as "brotherly help" (47%) or "defense of their own territory" (48%). Almost half (43%) believe that the Nazis are responsible for the Katyn massacre, which involved the murder of around 22,000 Polish citizens, including army and police officers, by the Soviets in the spring of 1940. Only 26% of Russians are aware of the Stalinist apparatus of repression's responsibility for this act (*Obraz Polski,* 2020, pp. 22–23).

The simple identification of disinformation may be considered an adequate defense mechanism if the recipient detects it, recognizes it as a threat to the cognitive process, and rejects it, rendering it permanently unreliable. Recognizing false or manipulated information is not a straightforward task, however; it requires theoretical knowledge and practical skills in critical thinking and fact-checking. These competencies, bolstered by open-source intelligence (OSINT) techniques, provide a broad range of possibilities for every disinformation analyst.

CHAPTER 7

Disinformation Analysis – First Line of Defense: Critical Thinking, Fact-Checking, and Open-Source Intelligence

1 Critical Thinking

The ability to separate facts from opinions and truth from lies and to recognize manipulated or false content is particularly important in times of post-truth. This competency can be acquired, developed, and improved through critical thinking, but it should not be equated with criticism in its common understanding. Its purpose is not to maliciously spot errors (as with criticism) but to skillfully analyze information and assess the accuracy and logic of argumentation. The term "critical" is derived from two Greek words comprising correct thinking based on intellectual standards: *kritikos* (luminous judgement) and *kriterion* (standards, criteria). Critical thinking is the reflection and analysis that accompanies decision-making and problem-solving. It is based on logical, careful inference and is a process guided by good evidence. It defines problems, identifies contradictory arguments, uses relevant and reliable data, asks fundamental questions, and effectively uses information to make assessments and appropriate decisions (DiYanni, 2016, pp. 17–18). Critical thinking practices are documented as far back as ancient times. Richard Paul and Linda Elder believe that "critical thinking is directly aimed at achieving a well-founded opinion, using adequate standards of evaluation that determine the true meaning or value of something" (Paul & Elder, 2014). Edward Glaser distinguishes its three components as:

1. An attitude of willingness to consider, in a deliberate manner, problems and objects that fall within the scope of experience.
2. Knowledge of logical methods of reasoning and inquiry.
3. Some practice in applying these methods (What is, 2021).

Critical thinking is therefore constructive, useful, and necessary. It is a conscious, well-thought-out, and orderly process. It includes analysis, evaluation, interpretation, and judgement. It is a practical set of skills such as: gathering necessary information; analyzing; determining significance, materiality, and reliability; using in practice; processing; and drawing conclusions.

The cyclical model of critical thinking has five components: (1) asking questions about particular knowledge; (2) clarifying what the knowledge means;

(3) seeking evidence for the knowledge; (4) making an assessment based on objective criteria; and (5) reflecting on assumptions and possibilities of utilizing the knowledge.

According to Robert DiYanny (2016), a key element of critical thinking is to ask the right questions and to give reasoned answers. This method is based on asking the questions: "How do I know what I think I know?" and "What evidence do I have for what I think I know?" Simply put, humans learn in three ways: first, empirically, through an individual experience of the world; second, by reading and listening to others; and third, by self-discovery. Critical minds constantly challenge both their own reasoning and the thought processes of others. Their attitude is marked by deliberate skepticism, rejecting the acceptance of unsupported assumptions (DiYanni, 2016, pp. 19–30).

According to Richard Paul and Linda Elder (2014), the main questions of critical thinking involve the following:

- What is the design and purpose of a specific thinking process?
- What question or problem requires reflection?
- What is the point of view or perspective?
- What thesis or views are developed and why?
- What facts, information, or data support the idea or thesis?
- What assumptions have been made, and which of them can be challenged or questioned?
- What conclusions can be drawn, and what is the conclusion from them?
- Are there implications and consequences that can be assumed?
- What concept or theory underlies such thinking?

Critical thinkers are characterized by *intellectual honesty*; they see facts as they are, not as they would like to see them. They are also characterized by their *humility*, or ability to perceive the complexity of the world and consider the processes taking place in it as complicated. They avoid arrogance and egocentrism, are aware of the limitations of their own knowledge and demonstrate *perseverance* with a consistent and methodical approach to problems and questions. Despite difficulties, they do not choose easy and quick solutions; instead, they have *courage* and a conviction that their own ideas and arguments are right even when they do not agree with the majority opinion. They also demonstrate *empathy*, appreciating different perspectives and opinions, as well as *autonomy*, with their ability to think independently and take responsibility for their evidence and arguments (Paul & Elder, 2009; DiYanni, 2016, p. 92).

By developing critical thinking skills, it is possible to better understand the reality that surrounds us. It improves our ability to deal with problems, see cause-and-effect relationships occurring in various phenomena and processes, assess the significance and usefulness of specific information, and better argue and defend specific positions (Henzler, 2018, p. 9).

Critical thinking translates into competence of separating "facts" from "pseudo-facts", or truths from falsehoods. Facts, data, and information require analysis and interpretation to be useful and to serve rational conduct. Placed in a false context or surrounded by false inferences, they become useless and harmful. They must therefore be confirmed and verified first to be considered reliable. Only then do they translate into real knowledge. It is worth asking the following four questions each time:

1. Where does the information come from, and what is its source?
2. What is the evidence, not the claims, for it?
3. Why is this information important, and what is its value?
4. Is it complete, and what is missing?

According to DiYanni (2016), each thesis can then be analyzed in detail by following the steps below:

- Dismantling: eliminating irrelevant details of formulations to better understand the main thesis and its essence.
- Following: finding out who says what, who supports a given thesis, and who opposes it, as well as what others think about it.
- Analysis: compiling and presenting reasons for considering a given statement to be true, as well as confronting it with the knowledge resulting from empirical experience and the evidence supporting it.
- Reflection: deciding to accept what is proposed, supported, or stated (pp. 52–53).

In this process, the knowledge and experience of the recipient of the information are greatly important. Some believe that learning and gathering knowledge in the digital age no longer makes sense, as "everything can be checked online". There is a hint of a provocative reason in such a statement, but it should serve to reflect on the need for changes in education in a manner that ensures both the acquisition of competences resulting from new social phenomena and the transfer of traditional knowledge. A failure of technology, caused by the weakness of infrastructure, software or hardware, for instance, could result in a lack of access to online resources. Apart from that, much of the content presented on the internet is simply erroneous and based on false or manipulated information, interpretations, and conclusions. Not having a method of error screening translates into vulnerability to manipulation and disinformation and living in a world of apparent knowledge. When browsing information online and in traditional media, one should be skeptical and pay attention to several elements characterizing manipulated content, including:

- Sensational titles or outrageous leads. A key impulse for spreading false information is that it is based on extreme emotions. The sensational promises of

the titles are often not reflected in the text, however. They take advantage of the laziness of recipients, satisfied with the loud headline and not reading further. The purpose here is only to attract clicks ("clickbait") that translate into ad revenue from the website on which they are published.
- Emotional language and photos. Disinformers aim to arouse extreme emotions (usually negative) in the form of indignation or dissatisfaction, which finds its outlet in emotionally marked and often offensive comments fueling the discussion in the form of trolling.
- A lack of sources cited for the information provided. This can also include references to an unknown source of which the credibility cannot be quickly assessed or one that is known to publish fake news.
- Linguistic errors. Incorrect spelling or loanwords from foreign languages may indicate that the content was prepared by a person without appropriate qualifications. It may also mean that the editorial office does not have basic quality control mechanisms for the published texts or that the authors are using automatic translators. Actions of this kind are characterized, for example, by the activity of Russian and Belarusian disinformation in Poland.

Caution should also be used regarding published content with sources characterized by:
- the dominance of clickbait, sensational headlines, titles, and leads;
- a large number of advertisements;
- a URL that looks suspicious and mimics known network addresses;
- missing or incomplete information on authors, editorial staff, management;
- hidden information about the website in the WHOIS register, making it impossible to check who the owner of a given domain is and to further assess its credibility;
- a domain registered in a country other than the one in the language of which it publishes, which may indicate that it is used to conduct disinformation activities commissioned by another country (Głowacka et al., 2019, pp. 6–9).

2 Verification of Information

When material or content posted on a website is questionable to a critical audience, security mechanisms are activated. Sensational reports, emotional language, an unknown author or source, controversial theses, or a lack of supporting evidence should bring skepticism against mis- or disinformation. Even if the data and expert opinions presented in the article seem solid, the possibility remains that it is fabricated and untrue. Its legitimacy can be determined by means of information verification and fact checking. This method confirms or refutes statements appearing in the media or online to remove errors, enable the dissemination of the text, or reject it if it is not possible to confirm its claims.

The methods used by professional fact-checkers are often reachable to every internet user. Unfortunately, the vast majority of readers do not check the authenticity of information in any way, as the data quoted in the first part of the book illustrates. This may result from a lack of basic knowledge about the available tools, insufficient practical skills, a lack of time, or laziness. To verify the information, the recipient can use checklists, models and schemes, and digital forensics tools. The measure of effectiveness in this case is persistence combined with creativity, systematic work, and developed safety reflexes. The information verification methodology is based on five pillars:

1. Provenance – What are the origins of the information? Is the article a primary or secondary source? Who is the real producer of the information, and who only intercepted or used it? This applies to official and unofficial accounts, profiles, and websites. Determining the origin of the information helps to understand the context and the publisher's motivation. By not verifying the original source of information, news agencies can erroneously duplicate fake news simply because other media have done so. An example of this is the "blue whale's story", a computer game developed in Russia which supposedly caused suicide among children. The disinformation spread around the world not only because it had not been checked but also because it was duplicated by thousands of press agencies, newspapers, TV programs, and internet portals.
2. Source of information – Who is the author or editor, and what is known about them? Is the source of information generally recognized as reliable, professional, and credible? Is the source of information complete? Is the author of the material an expert in the given field, or do they refer to the knowledge of experts? Are the experts referred to in the text true experts in the given field? An example of this is the multitude of pseudo-experts formulating controversial theses about COVID-19 and the validity and effectiveness of vaccinations against it. They often only seemingly have such qualifications (e.g., they are doctors whose specialties do not pertain to immunology or infectious diseases). One of many examples is Carrie Madej, an American osteopath who is famous for spreading anti-vaccine conspiracy theories about the pandemic and vaccinations. Madej was one of the first to suggest that COVID-19 vaccines "would induce changes in human DNA" (Goodman & Carmichael, 2020).
3. Date of publication – The date of publication informs the reader when the material was created and if it could be outdated or have changed. By

determining this information, the recipient should be able to determine whether the text concerns current affairs or whether its subject matter is artificially related to other past events. Such action may be aimed at distorting the viewer's assessment of current issues. An example of this is the false content circulated about alleged protests against pandemic restrictions in France in 2021. In this case, photos were used from the "yellow vest protests" in 2018 to fabricate the alleged proof of the events (Reuters Fact Check, 2021).

4. Location – Do the content, graphics, and video materials correspond to the location and context they relate to? An example of this type of manipulation are the photos of two men with Arab physiognomy smiling on the backdrop of the burning Notre Dame Cathedral in Paris 2019. These photos were used to spread the false information suggesting that Muslim fundamentalists were responsible for the fire (Kobla, 2019).
5. Motivation – What are the author's intentions and goals? Do they act for social good by publishing proven and thoroughly compiled content, or do they manipulate information to radicalize public debate with controversial and disingenuous articles? This can also be driven by commercial motivation, focused on advertising profits, which are the highest in the case of clickbait articles containing sensational and emotional information (Urbani, 2019).

Among the numerous models, methods, and tools used to verify the truthfulness of information, one of the most popular and effective is the CRAAP model. The analysis model developed by Meriam Library at California State University consists of five elements:

1. Currency of the information
   - hen was the information published?
   - Has the information, if not new, been updated?
   - Does the issue related to the information require up-to-date data, or is it possible to rely on older materials?
   - Do the links, if any, in the information work?
2. Relevance
   - Does the information relate to the subject or answer questions important to the recipient?
   - For whom was the information prepared, and what is the target group?
   - Is the information adequate to the recipient's needs? Is it too vague, too advanced, or too detailed?
   - Have other sources of information been checked before deciding to use this information?
   - Does the public admission to use a given source of information look serious?

3. Authority
   - Who is the author, publisher, source, or sponsor of the information?
   - What are the credentials of the author of the information? What organization, institution, or other entity is it affiliated with?
   - Is the author qualified to speak on the subject?
   - Are there details such as publisher name, e-mail address, etc., next to the information?
   - Does the website address say something about the author or sender (e.g., does the URL end with: ".com", ".edu", or ".gov")?
4. Accuracy
   - Where does the information come from?
   - Is the information supported by evidence?
   - Was the information peer-reviewed or cited?
   - Is it possible to confirm at least some of the information provided in another source or through use of self-knowledge?
   - Does the language or pronunciation of all information indicate impartiality, and is it devoid of any emotional tinge?
   - Are there any spelling, grammatical, or stylistic errors in the information?
5. Purpose
   - What was the information created for? Is it supposed to educate, inform, entertain, or persuade?
   - Did the author or the person financing the creation of the information clearly define its purpose?
   - Is the information describing facts, presenting an opinion, or perpetrating propaganda?
   - Does the point of view presented in the information appear objective?
   - Is the information biased? Does it take a specific position on issues related to politics, religion, or social issues? Does it present the perspective of only one institution or person?

The CRAAP model allows for detailed content analysis, but it may seem too elaborate and complex for the average internet user. By applying it, the recipient spends more time analyzing the article than simply reading its content. In a "network society", a few clicks can give access to hundreds or thousands of articles on a specific topic. The primary challenge, therefore, is to make the audience spend more time carefully selecting information sources or reflecting more thoroughly on the content.

Despite its complexity, the questions formulated in the CRAAP model have a great analytical value in the field of information verification methodology. By paying attention to the currency, relevance, authority, accuracy, and purpose of the information, the recipient develops self-protection mechanisms against manipulation.

The International Federation of Library Associations and Institutions (IFLA) has also proposed a more simplified procedure for dealing with information. It consists of eight activities that any recipient of information should perform:

1. Check the sources. Conduct a thorough analysis of the website as well as information about the editorial office and its mission. Check the domain owner, financing sources, and contact details.
2. Check the authors. Find information about their experience and qualifications that translate into their credibility. Verify if the author(s) exist or are fictional characters.
3. Check the date. Sharing old information as "up-to-date" is often intended to distort the image of reality and lead the recipient to misinterpret the situation.
4. Read more. In the digital era, recipients face an overload of information every day. Consequently, this often leads to opinions being formed solely based on provocative headlines. After reading an article in entirety, it may turn out that the sensational title or lead is not reflected in its content.
5. Pay attention to bias. A careful reading of the article should include an analysis of whether the text is biased. Did the author consider the arguments of each party? Is the issue presented in an objective and balanced manner? The recipient must also pay attention to whether the author's personal beliefs affect their perception of the information.
6. Check additional sources. Manipulated content is often based on specific data, expert statements, and sources that are generally accepted as reliable. However, the quoted "data" or "expert opinions" are not always correct. The recipient should therefore check whether the sources provided in the links refer to the given information and whether they present it in a manner that reflects the original author's intent.
7. Check for satire. Controversial and sensational messages may take the form of satire (e.g., memes), which should be treated as a form of entertainment and humorous commentary on reality. Information and theses contained in various forms of satire should be verified with information on the same subject available in proven and credible sources.
8. Ask the experts. In the era of social media, the distance between the recipients of information and its journalists and experts has been significantly shortened. This enables internet users to confront questionable theories and information with the knowledge of specialists who

can answer concerns and questions in private messages. In this way, it is also possible to verify whether the assessment of a given situation or phenomenon referred to by the author of a text was actually formulated by them (*How To Spot*, 2017).

The practical application of this pattern should be a sufficient and effective mechanism for protecting against counterfeiting and manipulation in the information space. Recipients who do not have the time or are unwilling to independently analyze content can also look to professional organizations that verify information. According to Duke Reporter's Lab, along with the growing scale of disinformation and fake news, the number of fact-checking organizations in the world increased from 44 in 2014 to 149 in 2018. The dissemination of false information regarding the COVID-19 pandemic contributed to the expansion of the international network of information verifiers to 341 entities operating in over 100 countries in 2021 (Stencel, Ryan & Luther, 2022). Most often, these entities are non-governmental organizations (NGOs) associated with international networks such as: FactCheck.org, PolitiFact, The Washington Post Fact-Checker, Full Fact, Snopes, Fact Check from Duke Reporters' Lab, SciCheck, FlackCheck, Media Bias/Fact Check, NPR FactCheck, Hoax Slayer, and AllSides. More than 100 of them were affiliated with the International Fact-Checking Network (IFCN) in 2022 (*Empowering*, 2022).

Fact-checking organizations employ people experienced in working in the media, including former journalists; analysts; experts in public relations, public affairs, and marketing; as well as students and volunteers. Their activities are financed from funds provided by NGOs, internet platforms, private donors, and state institutions. Fact-checking organizations verify information and use unified standards and similar methods of work. The overriding principle within their work is transparency, consisting in the full availability of recipients to the methodology and information on the activities of a given entity. The basic standards and principles are contained in the IFCN Code of Principles for fact-checking organizations (Orsek & Ozsoy, 2020). These are:

- Impartiality, honesty, and justice. The fact-checker should check the statements of politicians from various groups to an equal extent, refrain from taking a position on the issues they are examining, and verify the facts. Judgement and evaluation belong to recipients.
- Transparency of activities. This includes sources, financing, organizational structure, and methodology. Fact-checkers should reveal what data and documents analyses are based on.
- A clearly defined content modification and correction policy. Fact-checkers have an obligation to be open and honest about text modifications, making the recipient fully aware of any corrections that were made.

– Intent to improve the quality of public debate. This should be the main goal of fact-checking organizations. They cannot be guided by political or commercial motivation or any other factor which could affect their objectivity, reliability, and credibility.

Since 2018, the European Union has supported the activities of fact-checking organizations and organized workshops and conferences as part of an independent European network of fact-checkers. The EU also provides them with its own tools and data. To facilitate the detection of various types of disinformation threats, the European Commission has provided funding for the Social Observatory on Disinformation and Social Media Analysis (SOMA), which created a pilot IT infrastructure to support fact-checking. In addition, many research projects are funded from EU grants to analyze disinformation and develop new content verification tools (Ogrodowczyk et al., 2020, p. 22).

3 Open-Source Intelligence

Competences and skills in the field of critical thinking and fact-checking help to identify disinformation. However, to analyze the content of a message; the source of its preparation, methods, and scope of distribution; and the determination of the objectives of a given operation all require the effective use of open-source intelligence (OSINT) tools. OSINT is one of the working methods of specialized law enforcement agencies, armed forces, intelligence services, investigative journalists, analysts, and scientists. Public access to data from open sources and tools for their acquisition, combined with extensive training, make OSINT skills the basic work tool of the network society.

Knowledge that used to be reserved for specialized state structures and obtained by means of operational activities can often be gathered using a private computer and free internet tools. The example of the location of the U.S. Army secret bases in the Middle East and Africa exposed to the public based on data from sports applications used by soldiers for training illustrates how challenging it is to ensure operational security in the 21st century. Every user of a smartphone, computer, and other electronic device leaves behind digital traces that can be analyzed through OSINT. This is evidenced by the activities of the international investigative journalist group Bellingcat and its efforts to unmask Russian hybrid activities.

The usefulness of OSINT is not only related to the development of information and communication technologies (ICT). It is estimated that during the Cold War rivalry, only 20% of intelligence came from secret sources, while as much as 80% was obtained by analyzing open-source data (Williams & Blum, 2018, p. 5).

Intelligence agencies often overestimate the importance of information obtained with the use of operational methods. At the same time, they ignore the conclusions that can be drawn from a careful analysis of open sources such as media, publicly available documents, social media reports, and other sources (Stróżyk, 2020, pp. 18–19). In intelligence activities, OSINT primarily provides the service with a broad background of the problem and builds an appropriate context for its further analysis. Often, the information contained in open sources turns out to be so detailed and certain that the further use of more expensive methods of information gathering is not justified. OSINT also allows for verification of information previously collected by other methods, which undoubtedly has a positive effect on the quality of the final analytical product. Such verification can help identify errors made by an information gathering team or detect manipulation and disinformation used by foreign intelligence. The NATO Open-Source Intelligence Handbook distinguishes the following categories of information from open sources:

- Open-Source Data (OSD). Data in raw form from printed, radio, TV, or digital (internet) sources. This also includes photos, recordings, satellite imagery, written notes, etc.
- Open-Source Information (OSIF). Information consisting of aggregated data, typically as a result of an editing, filtering, validation, and presentation process. This includes press, literature, scientific studies, publicly available reports, analyses, documents, etc.
- Open-Source Intelligence (OSINT). Information obtained from publicly available sources that has been processed, analyzed, and disseminated to recipients to answer a specific question.
- Validated OSINT. Information that can be assigned a high degree of certainty. It may, for instance, have been created by an intelligence analysis specialist who also has access to information obtained through operational means (*NATO Open*, 2001, pp. 2–3).

Any analyst involved in monitoring disinformation should develop an individual database of sources that they intend to systematically observe and analyze. It will be the basis for collecting "raw data", or OSD, and "publicly available information", or OSIF, which will be subject to a process of analysis and verification. When creating such a database, it is first necessary to identify the priorities of the analyst/analytical team and what kind of disinformation will be analyzed. Is it national or international? Is it generated by state actors (Russia, China, Belarus, Iran, Turkey), non-state actors (terrorist organizations, radical extremist groups), or business entities (competing companies)? Will the selected analytical approach be comprehensive or sectoral (focused on the spheres of politics, security, defense, military, economy, energy, or social issues)? To whom is the analyzed disinformation addressed (state institutions, business, society)?

After prioritizing an analytical activity within the wider field of identifying disinformation, the analyst can then proceed to selecting information sources (press agencies, press titles, TV channels, websites, social media groups, opinion leaders, etc.) that will be the object of analytical interest. In this process, the knowledge and competences of the analyst or analytical team should be considered. The process of identifying and analyzing disinformation may consist of the following stages:

1. Obtaining information.
2. Evaluating the information source.
3. Evaluating the information.
4. Circumstances surrounding the acquisition of information.
5. Identifying the possible purpose of the disinformation.
6. Evaluating the consequences of the disinformation.
7. Assessing the vulnerability to disinformation.

The main advantages of using open-source intelligence is the speed and ease of obtaining information, as well as its typically low cost for analysis. However, this does not change the fact that an analyst using OSINT must obtain information from reliable and proven sources, make selections to identify the essential information, and verify details. Any piece of information must have certain desirable characteristics such as timeliness, reliability, completeness, sufficiency, relevance, accuracy, unambiguousness, and comprehensibility. These characteristics must demonstrate its quality and value, which translate into its credibility and usefulness in achieving goals and decision-making processes. Valuable and useful information allows for analysis and evaluation of the current state of affairs as well as forecasts for the future. It is helpful in making optimal decisions and taking action.

If false or manipulated information is detected, many tools can be used to further verify and analyze it, including:

- Internet search engines. These platforms, which include Google, DuckDuckGo, Bing, Entireweb, and Yandex, help to identify information for analysis. It is important for analysts to note that different search engines may display results in different orders. Therefore, using only one search engine limits users to the "filter bubbles".
- Tools for collecting statistical data, tracking trends, and distributing information online. These tools, such as Google Trends, allow analysts to check the popularity of a particular piece of information.

- Content aggregators. Aggregators such as RSS Reader, News Reader, News Aggregator, RSSOwl.org, and Google News enable the systematic acquisition of content from individual internet sources.
- Scientific text databases. Databases like Academia.edu, ResearchGate.net, SSRN.com, and Google Scholar enable comparison of internet content with data, facts, theories, and scientific knowledge.
- Social media analysis tools. These include Netlytic, Social Bearing, Follow The Hashtag, TweetDeck, and Who posted what?. They allow users to track trends on social media, identify the distribution of specific content and articles, and analyze word combinations (including hashtags' popularity) to increase the reach of publications.
- Tools for advanced image search and analysis. These tools, including Google Images, TinEye, FotoForensics, RevEye Reverse Image Search, and Jeffrey's Image Metadata Viewer, help to verify the authenticity of photos and graphics and help analysts identify possible modifications.
- Tools for analyzing video materials. These tools, such as YouTube Data Viewer, InVID Verification Plugin, and VLC, offer similar analytical value to graphic analysis tools. They are still insufficient to detect "deep fakes", however.
- Domain analysis. This helps analysts search for information about changed or deleted websites and includes tools like ViewDNS, WhoIs, DNSdumpster, The Internet Archive Wayback Machine, Reddit, 4Chan Search, and DomainEye. Domain analysis allows analysts to determine who is the owner of a given website and gain access to their deleted but archived versions.
- Tools for acquiring and verifying information about people and entities. Such tools include Spokeo, Email Checker, PIPL, and Facebook Graph Search and they help determine if a given profile/account on social media is real or fictitious.
- Tools for identifying bots and fake accounts. These make it possible to determine the probability of whether a given account is managed by a human or a computer based on an analysis of its activity or number of followers. Examples include Hoaxy, Bot Sentinel, and Twitonomy.
- Tools for organizing collected materials. These tools, such as Hunchly, Maltego, and Start.me, allow analysts to organize the methodology of their work and create a personalized database of sources and tools for analysis. They also help to collect data and show the relationship between the information obtained (Hassan & Hijazi; 2018; Bielska et al., 2020).

When verifying the authenticity of information sources, particular attention should be paid to the following elements:

- URL address and account name. Discrepancies in the URL and name, including suspicious numbers or symbols, may indicate that the account or website is fraudulent.
- Profile photos. Disinformers use stock photos or photos created by photo generators when creating fake accounts. The same photos are sometimes used by several accounts, which can be verified with the use of Google Images, TinEye, or FotoForensics.
- Account activity and number of followers. Accounts publishing posts or comments intensively and regularly (e.g., at certain times of day) should arouse suspicion. Bot accounts usually last no more than a few months and don't have many followers. At the same time, they follow many other users which should raise suspicion. Bot activity may also be indicated by a sudden increase in followers.
- Published content. Fake accounts often do not publish much of their own content, instead re-posting others' content to increase their reach. By copying text fragments into the search engine and using the command "phrase in quotation marks", it is possible to verify if the same text was already published. SocialBearing and Hoaxy are also useful tools for analyzing the distribution of disinformation.
- Metadata. By checking metadata, such as date or geolocation, analysts can verify if a photo or video corresponds to the article's content (Głowacka et al., 2019, p. 11).

After collecting materials containing false or manipulated information, analysts can proceed with a detailed analysis. The materials collected during the investigation should also be documented and archived due to the fact that content posted on websites can be quickly removed or modified. A well-documented procedure also proves the transparency of the analyst's activities and constitutes a security guarantee in case of defamation accusations. The results of the investigation should be processed and published in the form of an analysis or comment on a specially created website, blog, or social media profile.

This methodology of work is crucial for any analyst. It takes time to develop an appropriate analytical workshop that allows the recipient to effectively navigate through the maze of information, as its excess is always a challenge. Artificial intelligence algorithms can facilitate the task in terms of precise data and content search. They should be used cautiously, however, as they can pose a threat of operating within a filter bubble (Pariser, 2011; Sumpter, 2018).

One concern that arises with this personalization of online search results based on user preferences is that it can lead to a distortion of the image of the analyzed phenomenon due to the omission of other available sources of information. A researcher may receive data that is only consistent with their expectations, views, tastes, or opinions, leading to a distorted judgement of reality.

Another problem related to the automation of the information search and analysis process is the difficulty to verify it. It is not easy to verify the date on which information is placed on the network, which complicates the verification of its credibility and affects the timeliness and usefulness of the fact-checker's work. Information that appears online is copied many times, usually without proper citation of its original source. A fully documented verification of where and in what form information appeared for the first time is therefore often near impossible. This phenomenon is aptly referred to as the "echo effect" (Best & Cumming et al., 2007).

Analysts must be aware of cognitive limitations and the typical traps of simplified thinking, which may lead to analytical mistakes. Such mistakes can be avoided by learning how to recognize them – first and foremost by employing critical thinking and adhering to certain intellectual standards and proven principles of good conduct.

In an effort to outline such principles, Former CIA Directorate Intelligence (DI) Officer Frank Watanabe formulated 15 analytical rules. They are popular with intelligence analysts and strategic analysis practitioners, although they do not formally meet the criteria of a scientific method or procedure. They can, however, be treated as specific principles of "analytical hygiene", originating from institutions with a high organizational and corporate culture. These 15 golden rules of a CIA analyst are as follows:

1. Trust your judgement. This does not mean ignoring the different interpretations, conclusions, or points of view of other people (reviewers). However, the essence of this principle is to not allow yourself to be influenced by others too easily, particularly if they are your professional superior.
2. Be bold in your theses and don't be afraid to make mistakes. Restricting yourself to a general interpretation of data and formulating conclusions only corresponding to your own state of knowledge on a given topic will certainly protect you from the scrutiny of superiors. However, it is difficult to accurately predict the future in this way. An in-depth analysis supported by rigorous logical thinking (despite stumbles) can contribute to obtaining key advance information at the right time.
3. It is better to be mistaken than to be wrong. It is not easy for analysts to admit they are wrong. The ability to correct one's assessment of the situation after obtaining new data or catching a mistake avoids the mechanism of "commitment and consequence", which can have dire consequences.

4. Avoid "mirror imaging" at all costs. Not every entity thinks and acts in the same way. Culture, nationality, religion, and political preferences all matter. Assessing the situation solely through the prism of one's own system of values is one of the biggest analytical pitfalls. Just because something is logical and consistent for someone does not mean that another person or group views the same process in the same way.
5. Analysis is worthless if it is not disseminated. An analyst's knowledge is useless if it is not developed and communicated to decision-makers in a clear and effective manner.
6. Coordination is necessary, but do not settle for the least common denominator. The convergence of arguments presented by various analysts may seem desirable because it confirms the belief in the accuracy of the formulated assumptions and conclusions. However, it does not mean that the analysis should only focus on aspects which experts agree on. The emerging differences of opinion between analysts and coordinators should not be removed from the analysis. If the analyst believes that they are right, the discrepancies in assessment between them and their coordinator should be presented in the text (e.g., in the form of a footnote).
7. If everyone agrees, something is wrong. Full agreement in a group of experts is rare. If such a situation is not preceded by a stormy discussion, the alert should be raised. The fact that everyone agrees may de facto mean that everyone is wrong (group thinking syndrome).
8. Do not write everything you know, just what is most important. Analysts often try to flaunt specialist knowledge, inundating the recipient with detailed albeit redundant information. The analytical text is not meant to be an extensive study of the topic, however. It must contain the most important information from the reader's perspective, which must be properly received by the recipient.
9. The form is never more important than the content. Although analyses are characterized by a high level of stylistic and linguistic value, attention to the form of the text should not adversely affect its substantive content.
10. Actively collect the information you need. If information is lacking, actively search for new data and information.
11. Do not take editorial proofreading too seriously. If editorial changes do not interfere with the content, accept them gratefully. However, the edits must not change the substantive value of the text or distort the essence of the analysis.
12. Chat with other analysts. This allows analysts to create networks of personal contacts with whom they can formally and informally consult substantive issues and through whom they can subject theses to a critical

analysis before writing. Influenced by other specialists, analysts can also come up with innovative ideas or create working teams on a given topic.

13. Do not let your career take priority over your assignment. Proper execution of analytical tasks is more important than promotion. The analyst's job is to deliver the best analysis based on the information available. Sometimes they may contain theses or conclusions undesirable for the recipient or supervisor. The desire to keep a job or get a promotion should not take precedence over performing analytical work properly.
14. Being an analyst is not a popularity contest. Analyses do not have to reflect the views of the recipient but the actual state of affairs. The role of the analyst often comes down to telling decision-makers what they do not want to hear.
15. Do not take your work too seriously. There will always be more work than time to get it done. Failure to maintain proper proportions between work and family or your passions will lower the quality of the analyses prepared (Watanabe, 1997, pp. 45–47).

The above rules mainly apply to analysts working in the state administration, operating in a certain formalized, bureaucratic, and hierarchical structure. However, they can also be helpful for independent disinformation analysts who must pay special attention to the operational security of their activities. Any actor working to unmask disinformation should therefore assume that they may become the target of attacks by the entity whose information activities they disclose. Finnish journalist Jessikka Aro, who exposed Russian influence and disinformation operations, was most convinced of this. Her activities were met with a wave of internet hatred directed against her by pro-Russian trolls (Aro, 2020).

Basic OPSEC best practices include using a password manager and two-factor authentication; using anti-virus software; systematically updating operating systems, applications, and programs; regularly backing up files to an external drive or to the cloud; selectively sharing personal information on the web or limiting its visibility to others (e.g., on social media); protecting sensitive information relating to identity and activities; limiting digital footprints; avoiding the use of open Wi-Fi networks; using encryption keys and secure applications like communicators that automatically encrypt messages; and anonymizing activities by using a VPN or TOR browser.

For the network society, critical thinking and fact-checking is an elementary set of practical skills necessary for the secure use of information resources available in cyberspace. Advanced information search and analysis methods are basic tools for the work of journalists, analysts, scientists, and state officials. All internet users are exposed to daily manipulation and false information that aims to disturb their cognitive abilities, change their perception of a specific phenomenon, or encourage specific actions or behaviors. Technological possibilities and the specificity of the functioning of cyberspace allow the dissemination of false or manipulated information on a mass scale. It gives a significant advantage to disinformation authors focused on political or commercial goals. The possibility remains, however, that in the near future new technologies such as artificial intelligence will be equally effective in detecting and counteracting this phenomenon.

In the meantime, the current first line of defense on the front of information warfare are individual skills in the field of critical thinking, data verification, selection of information sources, and assessment of their credibility and reliability. Well-understood skepticism and developed control mechanisms reduce individuals' vulnerability to disinformation and increase social resilience.

CHAPTER 8

Militarization of Information in Russian Strategic Culture

Disinformation is deeply rooted in Russian history, mentality, and strategic culture. One of the earliest examples of this are Potemkin villages (1787), set up by the order of Prince Grigori Potemkin on Tsarina Catherine II'S ROUTE to the Crimea, captured four years earlier in the war with the Turks. This case being obvious disinformation, however, perhaps it should be seen more as a component of the author's court culture where the satisfaction of the ruler is a confirmation of his or her greatness. It therefore becomes an imperative for officials and a driving force behind their career promotion. It is also worth considering the value of this reflection for the analysis of the contemporary Russian elite's rationality. Tsarina's experts also wrote the *Protocols of the Knights of Zion*, which was a disgusting anti-Semitic libel about a global Jewish conspiracy, useful in the political games in Russia at the time and helpful in reversing social attention from her problems at home. To this day, the text is used by various racist organizations.

1 Disinformation as an Element of Russian Strategy

The word "disinformation" was invented by the Soviet secret services in the interwar period. Their own practices of information manipulation were simply given a foreign sound, suggesting an external origin. Based on a solid tsarist tradition, Moscow has long used disinformation as a weapon against its domestic and foreign enemies. Its impact, power, and popularity grew with the development of technological tools and knowledge about individual and group cognitive processes (Galeotti, 2019; Rid, 2020).

Disinformation has also played an important role in Russia's imperial strategy. In tsarist times, the secret police, the Okhrana, had a primary goal of eliminating political opposition, subversive groups, anarchism, and terrorism. It was also responsible for carrying out information and psychological operations.

The secrets of disinformation were developed to perfection in the times of Soviet Russia. This can be seen in the MOCR Trust operation conducted in the 1920s by the secret police. Creating a fictitious opposition organization (*Monarchiczieskoye Objedinieniye Centralnoy Rossii*, or MOCR), the Soviet services developed the possibility of eliminating white emigration and misleading foreign intelligence supporting it. In the following years, its scheme was repeatedly copied by the Soviet secret services. Until the start of World War II, 25–40 operational games based on the Trust model were carried out, while 185 were conducted in the years 1939–1945. During the Cold War, the KGB used this method to disinform the CIA by planting American double agents such as Yuri Nosenko (Świerczek, 2020, pp. 7–8).

The former head of the Romanian security service, General Ion Mihai Pacepa, stated that during the Cold War, for the secret services of the Soviet bloc, disinformation was a more important task than classic espionage. The communist disinformation machine employed more people than worked in the Soviet army and defense industry altogether. The intelligence organizations alone employed over a million officers as well as several million agents and secret collaborators. Additionally, there were a number of people working in the international disinformation organizations established by the KGB, including the World Peace Council, World Trade Union Federation, World Democratic Federation of Women, International Student Union, and World Federation of Democratic Youth (Pacepa & Rychlak, 2015, pp. 52–57). They were all involved in misleading the West or supporting these efforts. By falsifying reality, creating and perpetuating myths about NATO and the United States, undermining the authority of democratic state institutions and the Church, and discrediting politicians and other personalities important to public opinion, the USSR shaped a negative image of the free world and inspired revolutionary movements to fight against Western imperialism.

From 1954, the First Main Board of the State Security Committee (KGB) was responsible for external Soviet disinformation activities (Andrew & Gordijewski, 1997; Mitrokhin, 2002). Operations of this type were carried out by Service "A", which was responsible for the use of offensive actions aimed at spreading disinformation or acting in a subversive and destabilizing manner. The term combines various techniques used in operations to influence the international environment of Russia and to support the Kremlin's policy (Darczewska & Żochowski 2017, pp. 12–14; Rid, 2020). The instruments of active measures included informational and psychological activities, disinformation, *maskirovka* (masking), special propaganda, provocations, subversion, and sabotage. These instruments made it possible to influence the political environment, public opinion, and the internal security of other countries, enabling Moscow to pursue its strategic interests. During the Cold War, active measures were used primarily to export the communist revolution, propagate Marxist-Leninist ideology, destabilize the political systems of adversaries, and discredit oppositionists.

Intelligence work disclosed by Pacepa now reveals that the thaw in relations between the U.S. and Romania in the 1970s was a disinformation operation planned by the KGB and the Romanian External Intelligence Service (DIE). By creating the image of a communist openness to the West, moderate and Europeanized Nicolae Ceaușescu hoped to obtain American investments and financial support stimulating the country's economic development. This operation was a test, and if it had been successful, it was to be replicated in other countries of the Eastern Bloc.

Its success was to be replicated by Mikhail Gorbachev in the form of his policy of openness (*glasnost*). Communist propaganda presented the operation as increasing the transparency of public life, democratization, and information openness to the world. It was also supposed to be one of the elements of the USSR's reconstruction (*perestroika*) policy, for which Gorbachev received the Nobel Peace Prize. In reality, *glasnost* consisted of polishing the image of the ruler for internal and external use. According to Pacepa, this model was also copied by Vladimir Putin at the beginning of the 21st century (Pacepa & Rychlak, 2015, pp. 29–37).

Another example of political disinformation described by Pacepa is the KGB operation code-named "Dragon". It was based on the conviction of the world public opinion that the murder of President John F. Kennedy in 1963 by Lee Harvey Oswald was a conspiracy by far-right American government circles. What was overlooked, however, is that Oswald served in the U.S. Navy in Japan in the late 1950s and was recruited by the KGB on the basis of ideological motivation. The effectiveness of this disinformation is evidenced by numerous books containing conspiracy theories as well as the success of Oliver Stone's Oscar-winning film *JFK* (1991). The film depicts members of the American military-industrial complex involved in the murder of the U.S. president, including the CIA, FBI, Secret Service, the mafia, and Vice President Lyndon B. Johnson. According to a Gallup poll, two-thirds to three-quarters of Americans today believe that there was indeed a CIA plot to kill Kennedy (ibid., p. 224).

The end of the Cold War and the collapse of the Soviet Union meant that external offensive information activity was severely limited. At that time, the focus remained on internal problems, and more specifically the war in Chechnya. Following former KGB officer Vladimir Putin's rise to power in 1999, the use of active measures gained momentum once again. Since 2000, Russia has developed a doctrine of information security and applied it in practice as a support for military operations in Georgia (2008), Ukraine (since 2013), and Syria (2015–2019). The immediate reason for the development of the doctrine was the Russian negative experiences resulting from two Chechen wars (Rodgers & Lanoszka, 2021).

Information security issues were emphasized in the Russian Military Doctrine of 2014, the National Security Strategy of 2015, and the Information Security Doctrine of 2016, with the final two having been updated in 2021. The New Strategy is another document formalizing and reinforcing the anti-Western vector of Russian foreign policy, which has been evident since at least 2007. Its content confirms that the state's security policy is prepared for a systemic and comprehensive confrontation with the West. Russia considers NATO to be the main threat to its security, particularly with the prospect of its enlargement to include Georgia and Ukraine and to bring military infrastructure closer to Russian borders. Internally, however, Russia fears a color revolution inspired from the outside, with the aim of destabilizing national authority and changing power in the Kremlin (*Wajenna Doktrina*, 2015).

The term "color revolutions", which refers to mass social demonstrations against undemocratic authorities, was first used to characterize protests against Slobodan Miloševic in Serbia (2000). This was followed by the Rose Revolution in Georgia (2003), the Orange Revolution in Ukraine (2004), and the Tulip Revolution in Kyrgyzstan (2005). The term "color revolutions" is also used in connection with the mass protests that have taken place in recent years in most post-Soviet states as well as in the Middle East and North Africa, including the Arab Spring. According to the Kremlin, these events were inspired and steered by the West to weaken Russia's influence in the zone of its privileged interests. Russian narratives describe the West's support for democratic transformations as a form of "hybrid war", which sharply contrasts the Western understanding of the term attributed primarily to Russia's aggressive actions against Ukraine (Bouchet, 2016).

The Russian National Security Strategy of 2021 emphasizes the growing importance of ICT used to interfere in other states' internal affairs, undermine their sovereignty, violate their territorial integrity, and disturb international peace and security processes. The Russian strategy directly accuses foreign secret service agencies of carrying out cyber-attacks on Russian information resources, conducting reconnaissance and intelligence activities, and taking actions aimed at destabilizing the social and political situation in Russia. This way of presenting security threats perpetuates a "sieged fortress syndrome", which describes the Kremlin political and military elites' paranoid conviction that the West aims to strategically encircle Russia by creating military bases in its vicinity. Russian political and military elites believe that Western states are constantly plotting to overthrow the regime in Moscow through revolutionary processes or an armed coup.

This syndrome is one of the ideological and propaganda foundations of the Kremlin disinformation activities against NATO (Darczewska, 2018). It justifies

Russia's territorial expansion ambitions as part of operations needed to defend itself from NATO enlargement to the East, framing it as the creation of a buffer zone used to separate foreign military infrastructure. Russian adherents of geopolitics ignore the fact that the length of Russia's border with NATO states is only 1/16 of the length of all its borders (1,215 km out of more than 20,000 km). On the other hand, it enables the creation of a false image of NATO as an entity seeking expansion, conducting hostile actions, and destabilizing security in Europe.

The five most often repeated Russian myths about NATO are: (1) NATO is at war with Russia in Ukraine; (2) NATO promised Russia it would not expand after the Cold War; (3) NATO is aggressive and a threat to Russia; (4) NATO is encircling and trying to contain Russia; (5) NATO's interventions in the former Yugoslavia, Kosovo and Libya prove that the Alliance is not defensive (*NATO-Russia, 2022a*). Through these narratives, Russia effectively frames itself as a victim and justifies its offensive actions. This message is addressed both to external and internal audiences; Russia constructs its image by sending political signals to Western countries and a message to Russian citizens, calling on them to unite against an external enemy. Such actions help ensure support for Putin's authoritarian regime.

In its most recent 2021 Information Security Doctrine, Russia included information activities in its catalog of defensive and offensive military capabilities, pointing to the possibility of an outbreak of interstate conflict because of activities in cyberspace. In addition to this, Russia promotes the concept of "sovereign internet" (RuNet) and aims to increase its influence in the field of global regulation of network development (Legucka, 2021a). The concept of a new generation of warfare (often referred to as hybrid war, non-linear war, or Gerasimov's doctrine) proves Russia's attribution of the strategic role of weaponization of information. The chief of the Russian General Staff, General Valery Gerasimov, observed that war is now conducted by a roughly 4:1 ratio of nonmilitary and military measures. In his views, these non-military measures of warfare include economic sanctions, disruption of diplomatic ties, and political and diplomatic pressure. The Russians see information operations as a critical part of non-military measures. They have adapted from well-established Soviet techniques of subversion and destabilization for the age of the Internet and social media (Waltzman, 2017, pp. 3–4).

In a short text published in the niche military journal, the Russian chief of the General Staff characterized the actions by the West in the Middle East during 'Arab Spring' as hybrid war. Yet he never referred to hybrid war as a Russian type of warfare. He claimed that the West's aim was to gain new footholds through color revolutions in post-Soviet states. He went on to reason that

Russia was forced to respond to the West using the same methods the West was using due to the increasingly blurry line between peace and a state of war (Gerasimov, 2013). Russia's strategic interests are therefore to be achieved by non-military means and indirect methods such as:

- Disinforming political elites, military commanders, and the public by manipulating information, fabricating information, falsifying reality, and distracting attention from Russia's real actions and goals (*maskirovka*).
- Social maneuvering. This involves intentionally exerting influence on society in order to achieve specific benefits.
- Compromising, corrupting, and blackmailing political and military elites.
- Fueling internal and external tensions and disputes and supporting separatist tendencies as well as ethnic and religious conflicts.
- Organizing provocations, riots, and demonstrations using the potential of protest.
- Supporting opposition groups, resistance movements, and extremist circles; creating institutions, associations, foundations, organizations, and armed paramilitary groups controlled by special services.
- Inspiring events that destabilize the internal situation; conducting subversive, sabotage, and terrorist activities, the aim of which is to evoke a feeling of uncertainty and threat in the society (Bryjka, 2018, pp. 168–180).

The common denominator in these activities is their aim to create controlled chaos, which enables these actors to shape the socio-political situation outside Russia by inspiring crises and then imposing a solution favorable to the Kremlin (Galeotti, 2017). Due to the development of information technologies, one of the key environments for this type of operation is cyberspace (Harrel, 2015; Bryjka, 2015, pp. 115–131). The virtual world is a key element of influencing public opinion due to its accessibility and openness, which removes many traditional communication barriers. The effect of this is a significant expansion of the scale of operation by means of active measures.

In the information and psychological space, these operations are conducted by propaganda channels and state centers controlling them. The Russian ecosystem of disinformation includes:

- State institutions. This is the presidential administration, including spokesman Dmitry Peskov and Deputy Chief of Staff Alexei Gromov in particular; the Ministry of Foreign Affairs, with Sergey Lavrov and Maria Zakharova acting as central mouthpieces; and others.
- Media and news agencies. These include *RIA Novosti*, *TASS*, *Interfax*, *RT*, *Sputnik*, *Golos Rossii*, *Rubaltic*, *Politnavigator*, *Baltnews*, and *Ukraina.ru*.
- TV channels. Some of the most popular channels are *Rossija-24*, *Pervy Kanal*, *Rossija 1*, *TVZvezda*, *and NTV*, while well-known propagandist journalists

include Dmitry Kiselyov, Vladimir Solovyov, Margarita Simonian, Olga Skabeyeva, and Yevgeny Popov.
- State-controlled, non-governmental organizations. The most notable of which are Rossotrudnichestvo, the Russkiy Mir Foundation, Russian House, and the Russian Center for Science and Culture.
- Whistleblower portals. These include WikiLeaks and DCLeaks.
- Troll farms. The Internet Research Agency is among the most well-known.
- Various types of alternative media.
- Social movements. These include the Eurasian Youth Union of Igor Panarin and the International Eurasian Movement of Alexander Dugin.
- Think-tanks. Most notable among them are the Russian Institute for Strategic Research (RISI), the Council for Foreign and Defense Policy (SVOP), the Russian Council for International Affairs (RIAC), the Center for Strategy and Technology Analysis (CAST), the Center for Energy and Security Research (CENESS), and the Center Strategic Research (CSR), Katehon.
- Geopolitical websites. These include *New Eastern Outlook*, *Global Research*, *News Front*, *South Front*, and *Geopolitica.ru*.
- Websites belonging to pro-Russian organizations, associations, and foundations. Some of the most well-known ones include Russkiy Mir, Valdai Club, the Gorchakov Foundation, the Institute for Civilization Dialogue in Berlin, the Institute of Democracy and Cooperation in Paris. (Smagliy, 2018; Barbashin & Graef, 2019; *GEC Special Report*, 2020; Kuzichkin & Hanley, 2021).

Russian security structures are primarily responsible for activities in the information sphere. These include:
- The Department 'K' of the Ministry of the Interior, which is responsible for the control of information flow and combating crimes in cyberspace.
- The Federal Security Service (FSB), which works on counterintelligence, combating internal threats such as terrorism, and collecting tactical intelligence, mainly in countries neighboring Russia. Within the structure of the FSB, there is an Information Security Center responsible for monitoring the internet and detecting information activities that threaten Russia's security.
- The Foreign Intelligence Service (SVR), which carries out external intelligence activities. The SVR plays a key role in Russian influence operations directed against NATO and EU member states. In the media sphere, SVR directs the work of *RT* and *Sputnik*. The main units responsible for developing guidelines for information content are the Analysis and Information Department and the Directorate of Foreign Intelligence.
- The Main Intelligence Directorate of the General Staff (GRU). This Russian military intelligence, which has the unit number 54777, is responsible for information and psychological activities. It was involved in interfering

with the presidential elections in the U.S. in 2016, and it works closely with Prigozhin's troll factory, controlling over 1,300 people and websites (*The GRU*, 2022).

The FSB and SVR, which were established after the collapse of the USSR, are the heirs to the KGB. Although in official nomenclature "active measures" have been replaced with the term "support measures" (*meropriyatiya sodeistviya*), their methods of operation have not changed significantly. Their activities, however, have extended to new technological possibilities. Foreign influence operations, for instance, fall primarily within the competence of the Russian SVR and GRU, which have a full spectrum of subversive methods, including disinformation, cyber-operations, sabotage, and subversion (Radin et al., 2020). Their activities may be supported or secured by the FSB. Within their influence operations, the SVR focuses on political, economic and social matters, while the GRU is the leading institution in military affairs. However, this does not mean that military intelligence does not participate in operations aimed at influencing the spheres of politics, the economy, or society (DiResta & Grossman, 2019). The best illustration of this intelligence cooperation between the FSB, SVR, and the GRU is the involvement of these institutions in electoral interference.

2 Russian Interference in Democratic Elections

In democratic states, the process of electing someone to a position of power by society is essential for ensuring the legitimacy of the entire political system. Any external interference aimed at disturbing or influencing the results of an election violate the law and international rules. Despite this, FSB Colonel Sergey Rastorguev has openly stated that he believes interference in the political processes of foreign states is an element of information warfare that can be used to achieve geopolitical goals. According to Rastorguev, the essence of this type of operations is "taking control of states and nations by introducing into the ranks of the leaders of a foreign country their own supporters or adherents of their own ideology and interests solely by means of the methods of informational influence". Rastorguev has further emphasized that "only a naive person can believe that the election of a president is the work of its inhabitants themselves and is not related to geopolitics". For this purpose, Russian intelligence operatives use many methods and techniques known as "dirty election technologies" (Wojnowski, 2021, pp. 6–8). Their goal is to influence the internal situation of the country where the elections are being held and to actively support a preferred candidate or political group.

Russian electoral interference is not a new phenomenon. According to Dova H. Levin (2020), from 1946–2000, as many as 74% of election interference efforts were directed against NATO members (p. 155). Since 2004, this phenomenon has been identified in 27 countries (Wojnowski, 2021, pp. 9–10). From November 8, 2016, to April 30, 2019, Russian actions were recorded in 16 out of 20 cases of interference in elections worldwide, with the rest attributed to China, Iran, and Venezuela (Hanson et al., 2019).

In 2016, the Russian intelligence services intervened in the U.S. presidential election, securing an easier path to victory for Donald Trump. A five-volume analysis prepared under the auspices of the US Senate Intelligence Committee (Russia Active Measures … 2020) provides an in-depth analysis of Russian information operations during this election. In the 2016 Brexit referendum, Russian active measures supported the campaign for Britain to leave the EU, thereby deepening divisions and crises within the community. During the presidential elections in France in 2017, Russia sided with the far-right, anti-European, and pro-Russian Marine Le Pen. In the same year, they supported separatism in Spanish Catalonia and the anti-immigration, nationalist Eurosceptical political Alternative for Germany (AfD) party during parliamentary elections (Applebaum et al., 2017).

Russian interference in democratic politics includes a wide range of activities, including cyber-attacks on political parties and governmental structures; the hacking of private and business emails followed by controlled leaks of stolen data (*hack-and-leak*); and mass dissemination on social media using bots, trolls, propaganda channels, networks of pro-Russian think-tanks, and agents of influence. The aim of these activities is to exacerbate existing social tensions, undermine citizens' trust in democratic institutions, promote people and political groups that are friendly to Russia, and discredit their opponents, thereby creating an atmosphere of chaos and uncertainty (Davis, 2018; Kruglashov & Shvydiuk, 2020, pp. 79–93).

One of the largest-scale examples of Russian interference was revealed in the case of the U.S presidential election in 2016, in which the Russian intelligence services supported Donald Trump as a candidate considered more favorable for the Kremlin by attacking his opponent, Hillary Clinton (Menkiszak, 2016). According to U.S. investigators, the decision was made by the President of the Russian Federation himself. The effort was a part of cyber-operations carried out by hackers from the APT-28 group (controlled by the FSB and SVR) and APT-29 (GRU). Russian intelligence services obtained e-mails from a private server and made them public on the whistleblowing portals WikiLeaks and DCleaks to compromise Clinton's campaign (Bassat & Cohen, 2019).

The distribution of the stolen materials compromising Clinton was supported by *Sputnik*, *RT*, Yevgeny Prigozhin's troll factory (the IRA), agents of influence, fake entities like Guccifer 2.0, and a network of fake websites and social media accounts created by the GRU specifically for the purposes of this operation. According to U.S. intelligence, the IRA financed over 3,500 ads that reached over 11.4 million Americans and created 470 pages on Facebook, on which over 80,000 posts were published with a reach of up to 126 million views (*Exposing Russia's*, 2018). The Russian operation of influence against the U.S. consisted of several stages:

- Segmentation of society. This includes selecting socio-political groups/fractions at which the information and psychological impact is directed. For instance, these groups could be white Americans of Anglo-Saxon origin (WASPs), African Americans, Latin Americans, Muslims, or LGBTQI+ community members.
- Polarization of socio-political groups. This involves creating or supporting internal divisions based on key political, social, and economic issues. For example, this could include highlighting social inequalities among African Americans and Latin Americans or inciting conflict between groups like white supremacists and LGBTQI+ persons.
- Initiating the political activity of a given group. This was done by transferring informational activities carried out in the virtual world into reality through rallies, demonstrations, protests, and riots. One example of this was the Heart of Texas group formed by the IRA, which organized the demonstration "Stop Texas Islamization" by a group of armed supporters. It also included Muslim counterdemonstrators organized by the United Muslims of America, another group founded by trolls.
- Controlling chaos. This stage involved the gradual escalation of social and political conflicts through information activities and possible corrections of the original operational plan (Wojnowski, 2021, pp. 15–19).

The disinformation campaigns led by the GRU through the IRA were aimed at inflaming public debates in the U.S. and playing on the high levels of political polarization of American society. In turn, there was a mutual reinforcement of the phenomenon as the polarization fostered the effectiveness of the disinformation and the effective disinformation strengthened the polarization (Misiuna, 2021, p. 63).

The use of the hack-and-leak method directly targeted Clinton's image, lowering her ratings and ultimately contributing to her defeat. The first piece of information about possible external interference appeared shortly after the elections, prompting President Barack Obama to initiate an investigation as he was leaving office. A joint statement by the Office of the Director of National Intelligence (ODNI) and the Department of Homeland Security

(DHS) confirmed the participation of Russian secret services in cyberattacks on servers of the Democratic Election Committee, the Clinton Foundation, election infrastructure, and several other institutions. This led to the imposition of sanctions on Russian intelligence officers and the companies that provided them with software or hardware, as well as the expulsion of 35 Russian diplomats from the U.S. (Piotrowski, 2017). Russian interference in the elections was also confirmed by the FBI, several private cyber intelligence companies, the National Intelligence Council (NCI), which brings together all 16 American intelligence institutions, and the investigation by special prosecutor Robert Mueller (*Report on The Investigation,* 2019).

During the 2020 presidential elections in the U.S., Russia tried once again to influence the public debate, working to deepen divisions on racial, political, and religious grounds. Russian intelligence activity focused on inspiring tensions between racial minorities by creating false accounts of the Black Lives Matter movement on social media while also organizing counterdemonstrations of their opponents under the banner of Blue Lives Matter. False information was also spread to discourage black voters from voting. In contrast, far-right circles were encouraged to spread hate speech. The 2020 operation was not conducted directly by Prigozhin's troll factory as it had been in 2016 but through a proxy organization acting on Kremlin's behalf, the Eliminating Barriers for the Liberation of Africa (EBLA) organization, founded in Ghana in 2019 (Wojnowski, 2021, p. 20). Apart from Russia, China and Iran were also revealed as actors who attempted to influence the results of the U.S. presidential election in 2020.

3 Russian Military Disinformation

In the Russian armed forces, the following entities are responsible for information and psychological activities:

- At the strategic level: The Information and Foreign Communication Center within the Ministry of Defense and the Special Action Center of the General Staff of the Russian Federation.
- At the operational level: The headquarters of the military districts (West, South, North, Central, and East).
- At the tactical level: Groups of psychological operations within independent brigades (e.g., the 130 Motorized Brigade), special forces (e.g., Spetsnaz), and the airborne regiments (45 Airborne Regiment). They have radio stations and TV studios, mobile printing sets, and unmanned aerial vehicles (UAV), among others. There are also specialized units responsible for PsyOps in the

Russian armed forces, including Unit 03126 stationed near St. Petersburg (Weiss, 2020, pp. 44–48).

The 12th GRU Information Operations Directorate also conducts military information and psychological operations (Bowen, 2021, p. 4). In the Russian military intelligence, there is a special unit 54777 (the 72nd Special Service Center) that was developed during the First Chechen War (Weiss, 2020). The unit is responsible for disinformation campaigns and played a leading role in spreading false information about the COVID-19 pandemic (Barnes, Sanger 2020).

The GRU structures also include other units that carry out activities classified as active measures. These include:

- Unit 29155, which is responsible for subversion and sabotage. This includes the operation to seize Crimea in 2014, bomb attacks on ammunition depots in Czechia in 2014, the murder of the Bulgarian arms dealer Emilian Gebrew in 2015, preparation of a coup in Montenegro in 2016 intended to block the country's membership to NATO, and the attempt to poison former GRU officer and British MI-5 agent Sergei Skripal in 2018.
- Units 26165 (APT-28; also known as Fancy Bear) and 74455 (Main Special Technology Center alias Sandworm), both under the GRU 6th Directorate of Signal/Electronic Intelligence, are responsible for cyber operations. These include the hacker attacks carried out during the 2016 and 2020 presidential elections in the U.S., the presidential elections in France in 2017, and parliamentary elections in Germany in 2017. They were also responsible for mass attacks using NotPetya malware, resulting in losses for hundreds of companies and institutions to the tune of USD 10 billion. Additionally, these units perpetrated cyber-attacks on the World Anti-Doping Agency and planned attacks on the Organization for the Prohibition of Chemical Weapons in 2018 (Bowen, 2021, pp. 11–18).

Military disinformation is firmly rooted in the Russian tradition and practice of special operations. Strategic disinformation was used in 1955, for instance, when the Soviet organization conducted Aviation Day, during which countless squadrons of new long-range M-4 bombers flew over Moscow. The display was coordinated with KGB-controlled leaks of secret documents indicating that the Soviet Union had not only overtaken the U.S. in the number of long-range reconnaissance aircraft (Tu-20) and strategic bombers (M-4), but had also attained better technical characteristics than their American counterparts (B-47 and B-52). In reality, however, only one M-4 squadron flew in the demonstration, appearing in the sky every few minutes because Moscow did not have more aircrafts at its disposal.

A similar example of disinformation can be seen in the Soviet operation code-named Walnut II. Its purpose was to convince the U.S. of a Soviet missile advantage. Initially, the CIA was deceived by these manipulations; but in 1957, when U-2 reconnaissance missions began, the actual state of the Soviet air fleet and missile potential was revealed (Pacepa & Rychlak, 2015, pp. 218–219).

Operation Anadyr serves as an example of Soviet disinformation that occurred at the operational and tactical level. Its aim was to conceal the deployment of Soviet missiles with nuclear warheads in Cuba. To mask the transfer and deployment of military personnel, weapons, equipment, ships, and airplanes belonging to various countries of the Eastern Bloc were used as the Soviets tried to prevent the cargo and its destination from being identified. However, the operation was uncovered by U.S. U-2 reconnaissance flights and led to the Cuban crisis in 1962 (Gribkov & Smith, 1993).

A more modern example of Russia's use of military disinformation was its Kavkaz military exercises, used in 2008 to prepare for its invasion of Georgia. To block Georgia's aspiration for NATO membership, Russia played on the separatist tendencies in two of its regions: Abkhazia and South Ossetia. As a result of Russian "encouragement", South Ossetian troops carried out provocations through artillery shells on Georgian villages to force a military response from Tbilisi and give Moscow a pretext to invade. Moscow used the Kavkaz 2008 military exercises to transfer weapons to separatist regions and increase the presence of Russian troops. During the exercises, it transferred approximately 40,000 additional soldiers to Abkhazia, the Pankisi Gorge, and bases in Ugardant and Java. It also increased the activity of its Black Sea Fleet, which clearly indicated preparations for the invasion. President Mikheil Saakashvili decided to defend Georgian territory, and Georgia's armed response to South Ossetian provocations gave the Kremlin the pretext it needed to launch an on-land offensive on August 8, 2008. The official cause given was "protection of Russian citizens" (Asmus, 2010, p. 286).

Russia used the Kavkaz, as it did with other military maneuvers, including Vostok (East), Centr (Center), Grom (nuclear exercises), and Zapad (West) for strategic communication purposes (Ratsiborynska et al., 2020: Petraitis, et al., 2021). Russia's neighbors have often been subjected to psychological and military pressure, which has been used to force them to pursue a security policy in line with the Kremlin's interests, including coercing them into refraining from joining alliances to which Russia is not a party. Moscow recognizes the "countries of the near abroad" for their zone of privileged interests and sees them as a buffer zone separating them from NATO. By demonstrating its strength, Russia not only shows its military capabilities but it also signals its readiness to use those capabilities to ensure its goals and strategic interests – despite the fact that these actions violate the fundamental principles of international law established after 1945.

Russian military maneuvers are not transparent, and they violate the provisions of the Vienna Document on Confidence and Security Building Measures (CSBM). The document obliges member states to notify the OSCE of temporary relocations of units and to invite foreign observers to the exercise. Members are required to give notice, for example, in any military situation where the number of participants exceeds 13,000 soldiers (*Vienna Document*, 2011). To get around these requirements, Russian military authorities deliberately split its military exercises, which in fact often include more than 100,000 soldiers. It is then able to conduct several smaller but fully integrated exercises at the same time while bypassing the limits. In addition to Moscow violating its own international obligations, such practices deliberately mislead Western states and prove Russia's unwillingness to build trust and minimize the risk of undesirable incidents.

The alleged defensive nature of such exercises is a pure form of disinformation. Russian-Belarusian Zapad maneuvers serve as the best example of this. In 2009, they simulated nuclear strikes against Warsaw. Since 2017, the exercise's scenarios suggest that NATO countries (Poland and Lithuania) support armed separatism in Belarus, which is what caused the direct military confrontation between NATO and the State Union of Belarus and Russia.

In 2021, this was practiced in Belarus in relation to the massive public protests against Aleksandr Lukaszenko's election fraud in August 2020. The Belarusian regime, with the support of Russian propaganda, directly accused Poland, Lithuania, Ukraine, and the U.S. of inciting a color revolution, destabilizing the internal situation, using hybrid methods, supporting separatist tendencies, organizing a *coup d'état*, and even preparing for NATO aggression.

The Zapad-21 exercises were accompanied by a disinformation campaign intended to present them as defensive action by Belarus and Russia against the growing risk of NATO's aggression, when in fact the maneuvers were offensive in nature (Bryjka, 2021a, pp. 157–180). On the one hand, Moscow has treated the Zapad exercises as an element of intimidation of public opinion in the countries directly bordering Belarus and Russia. On the other hand, Moscow has also used them as a tool to deepen divisions between the members of the transatlantic community by gradually disavowing information about the scale and nature of the threat. The psychological aspects of this campaign are also strategic, presenting Belarus and Russia as constantly threatened by the hostile policy of the West, which aims to destabilize them and overthrow the ruling regimes. This narrative justifies the use of adequate measures, including military operations, to counteract aggression (Wilk & Żochowski, 2021, pp. 4–6).

Keeping these practices in mind, the world observed the concentration of Russian troops on the Ukrainian border in the autumn of 2021, as approximately 150,000–200,000 soldiers were deployed with heavy military equipment. These actions were initially an element of political and psychological pressure directed against Ukraine, the U.S., and NATO, to which Moscow issued demands questioning the foundations of the international order and security in Europe after the Cold War. Violating the principles of the UN Charter, the OSCE Paris Charter, disarmament agreements and the NATO-Russia Founding Act, Moscow was sending Western decision-makers the message to the that it was not interested in negotiating peace but only in imposing its political will. However, Russia also did not rule out the risk of the conflict escalating or even of a full-scale invasion of Ukraine. The tense atmosphere was exacerbated by false allegations by Russian Defense Minister General Sergey Shoygu, who accused the West of preparing for a provocation in the Donbas with the use of chemical weapons (*American mercenaries*, 2021). Subsequently, U.S. intelligence made the information public about the provocations staged by the Kremlin, which they noted could be used to justify a full-scale invasion of Ukraine. According to CNN, Moscow transferred special forces to Donbas with the aim of conducting sabotage operations against pro-Russian separatists and the Russian-speaking population in Ukraine. Although rebels from the self-proclaimed Donetsk People's Republic (DRL) and the Luhansk People's Republic (LPR) have played the role of Kremlin proxy forces in a hybrid war in Ukraine since 2014, an attack on them by Russian special forces under the pretense that the actions were carried out by Ukrainian forces or a NATO state provided a pretext for the invasion of Kiev (Bertrand & Herb, 2022). In this case, the choice of U.S. intelligence agencies to release this top-secret information aimed to attribute the false flag operations carried out by Moscow and discredit its narrative.

Despite Western diplomatic efforts to de-escalate the situation around Ukraine, Russia's actions did not lead to easing tensions but instead to an escalation of the situation. This was evidenced by the cyber-attacks on Ukraine, suggesting cause for the deployment of Russian military systems in Cuba or Venezuela, and the transfer of Russian soldiers to Belarus under the official justification that they were "to participate in joint exercises" (Dyner & Kacprzyk, 2022).

On February 24, 2022, the world became fully aware that Russia's actions in Ukraine were not only an element of blackmailing, psychological pressure, or negotiation tactics but also preparations for a full-scale invasion. To justify the aggression, Russia formulated false accusations against Kiev regarding genocide in Donbas. This then created a pretext for the unilateral recognition of the independence of the self-proclaimed separatist republics and the launch of a "special military operation in Ukraine", as Russian propaganda describes the military aggression (Bryjka, 2022).

Similar actions, although on a much smaller scale, preceded Russia's aggression against Ukraine in 2014. At that time, Russia gathered approximately 40,000 soldiers on the border with Ukraine and had about 20,000 more stationed at the bases of the Black Sea Fleet. The intensive informational and psychological activities carried out at that time laid the ground for two special operations: (1) the occupation of Crimea by Russian soldiers deprived of identification marks, known as "little green men"; and (2) the organization of a pro-Russian rebellion in Donbas by the intelligence and special forces.

These activities were preceded by long-term disinformation and propaganda campaigns carried out since the Orange Revolution (2005) that had significantly intensified during the Revolution of Dignity, otherwise known as Euromaidan (2013–2014). The goal of these propaganda efforts was to strengthen pro-Russian attitudes by: promoting the Russian language and culture; emphasizing the historical ties of the Crimean Peninsula with Russia; distributing Russian passports; and circulating narratives that aimed to delegitimize the legal power of Crimea's accession into Ukraine in 1954. At the same time, anti-Ukrainian sentiment was inspired by disseminating false information about violations of rights of the Russian-speaking population in Ukraine and by anti-Western sentiment accusing the West of organizing a color revolution.

In Odessa, Kharkiv, Donetsk, and Luhansk, the Russian special services organized anti-revolution protests (anti-Maidans), which aimed to serve as evidence that some regions did not support Kiev's pro-Western aspirations. In reality, these actions also turned out to be part of preparations for armed aggression in eastern Ukraine (Bryjka, 2016, pp. 201–219). At the end of 2021 and the beginning of 2022, Russia intensified its anti-Ukrainian and anti-Western narratives disseminated for domestic audiences to prepare its own society for the invasion of Ukraine. It continued to perpetuate myths suggesting that Ukraine is an artificial state devoid of historical foundations, arguing that its existence as a state is merely a side effect of wrong decisions made by Soviet leaders. This message was then used to rhetorically undermine the right of the Ukrainian people to sovereignty and independence and to justify Russian claims against the lands inhabited by Ukrainians. Ukraine was also presented as a failed state, unable to protect its own citizens and deprived of opportunities for integration with NATO and EU structures. The Ukrainian authorities, in turn, were described as violating the rights of the Russian-speaking population, labeled a Nazi junta who came to power as a result of a Western-inspired color revolution (Bryjka, 2022).

Such rhetoric was used to justify the three main goals of the attack: (1) de-Nazification, (2) demilitarization, and (3) neutral status for Ukraine (*Putin's declaration*, 2022). It found fertile ground in Russian society, as confirmed by the research carried out by the Levada Center, which showed that 60% of Russians believe the U.S. and NATO are responsible for escalating the conflict in eastern Ukraine. The effectiveness of Russian war time disinformation has been further strengthened by the blocking of independent channels of information like Western social media and imposing draconian penalties, such as 15 years imprisonment for proclaiming "untruths" about the Russian "special military operation" in Ukraine. At the request of the prosecutor general's office, Roskomnadzor ordered publishers to remove the terms "intervention" and "war" from materials describing the situation in Ukraine, under the threat of a fine or the removal of their content. Additionally, media outlets were obliged to report on the situation in Ukraine only based on information from official sources (Legucka, 2022).

4 Disinformation as a Science and Practice of Russian Politics

Russians treat disinformation as a field of science and an element of domestic and international politics. The Russian theory of information wars is derived directly from the theory of special propaganda, which was taught as a separate subject from 1942 at the Military Institute of Foreign Languages. In the 1990s, special propaganda disappeared from the curricula as a consequence of the end of the Cold War rivalry between superpowers. After Vladimir Putin came to power, however, disinformation activities returned to the arsenal of the Kremlin's policy tools. In 2000, the Information Security Doctrine of the Russian Federation was adopted, which entrusted the work of information security to the Military Information and Foreign Languages Department of the Military Institute of the Russian Ministry of Defense. After the reorganization of the university, the training of specialists in the fields of organization of foreign military information and communication, information analysis, and monitoring and elaboration of military information resumed. Research in the field of special propaganda is also carried out by, among others:

- Moscow State Institute of International Relations (MGIMO).
- The Diplomatic Academy of the MFA of the Russian Federation.
- The Institute of Information Security Problems and Information-Analytical Center for Research on Socio-Political Processes in the Post-Soviet Area at Moscow State University Lomonosov.
- The Institute of Cryptography, Telecommunications, and Computer Science at the Federal Security Service (FSB).

- The State Science and Research Experimental Institute of Technical Information Protection Problems of the Federal Service for Technical and Export Control (FSTEC).
- The Federal Protective Service (FSO) Academy in Orel.
- The Voronezh Research Institute of Telecommunications.
- The Academy of the Russian Internal Affairs Ministry in Volgograd and Rostov-on-Don.
- The Scientific and Methodological Association of Higher Education Facilities of the Russian Federation Covering Information Security (Darczewska, 2014, pp. 9–10).

The current Russian theory of information wars was developed mainly by Igor Panarin and Aleksandr Dugin. Both are academic lecturers and representatives of the Russian geopolitical school. They describe Russia's offensive actions against Georgia and Ukraine as a defense against Western attempts to dismantle Russian statehood and its zone of privileged interests in the former USSR area. Furthermore, Panarin and Dugin are not only theorists but also practitioners of information warfare. They actively participate in the journalistic and analytical programs of *Kanal 1*, *Rossija*, *NTW*, *Ren-TV*, and *TV RT*. They and their concepts are associated with the Russian intelligence services.

The Russian model of influence operations, according to Igor Panarin, involves the following elements:

- Methods and techniques: These include social control, or the influencing of society; social maneuvering, or the intentional influencing of society to achieve specific benefits; information manipulation, or the use of real information in a way that causes false implications; disinformation, or the dissemination of manipulated or fabricated information; fabrication of information, or the creation of false information; lobbying; blackmailing; and extortion.
- Overt and covert tools: These include black, gray, and white propaganda; intelligence services that collect information about the enemy; analytical components like media monitoring and current situational analysis; an organizational component that coordinates and controls channels; and other interconnected channels, including the forces of special operations through subversive and false flag operations.
- Operational stages: This involves forecasting and planning, organization and stimulation, garnering feedback, correcting operations, and performance control (Darczewska, 2014, pp. 15–17).

After several years of Russian disinformation monitoring, the EU task force East StratCom, which has been in operation since 2015, identified a model of patterns of informational and psychological warfare. By obstructing the exchange of views based on disagreement with mutual respect, trying to dominate the debate, and imposing a specific narrative, pro-Kremlin trolls try to disrupt the possibility of real public engagement and dialogue. They try to manage societies by inflaming sensitive issues, fueling emotions, and arousing fears and social phobias. The tools of Russian disinformation defined by the EU vs. Disinfo team as modus trollelandi, or SWAMPED, includes techniques such as:

- Strawman fallacy. This involves attacking opponents for views they never expressed. Examples of this include unjustified and unproven attribution of fascist/Nazi sympathies to those who criticize Kremlin policy, accusations of Russophobia, or allegations of acting as an agent of foreign influence.
- Whataboutism. This is a technique of responding to an accusation or a difficult question by counter-accusing or raising another issue. For instance, authoritarian leaders may dismiss accusations of brutal suppression of protests, like in Belarus in 2020, by redirecting the subject and asking: "and why don't you deal with the yellow vests in France or the post-election protests in the U.S.?"
- Attacks. By using insulting language, like calling dissidents of the Kremlin media fascists, extremists, Satanists, perverts, and pedophiles, disinformers try to discourage their opponents.
- Mockery. This includes the use of sarcasm to undermine an opponent. Foreign Ministry Spokeswoman Maria Zakharova uses this technique in her weekly speeches, employing ironic remarks about "our Western partners" and "our esteemed colleagues" in a degrading manner. A similar approach was used at the January 2021 demonstrations in Russia when the pro-Kremlin media labelled thousands of protesters "Navalny's puppies and mama-boy revolutionists".
- Provocations. This includes using charges that are both provocative and baseless; for instance, provocations were used when the Kremlin accused Western countries of preparing to use chemical weapons in Syria and Ukraine.
- Exhaustion. This involves flooding the opponent with details and technicalities. This method was used, for example, in connection with the act of air piracy (hijacking a civilian airplane to arrest his political opponent Roman Protasevich) committed by the President of Belarus, Alexander Lukashenka. The pro-Kremlin media compared this to a 2013 incident in which the American authorities demanded that the plane carrying Bolivian president Evo Morales land in Vienna. This false equivocation continues to

be perpetrated while ignoring the fact that the 2013 case involved no false bomb alarms or interception by fighter jets.

- Denial. This involves questioning any evidence of the existence of disinformation. Examples of this include questioning the evidence of the presence of Russian troops in Crimea and the Donbas; questioning the results of the international investigation into the downing of the MH-17 plane; and questioning the evidence of the participation of GRU officers in the poisoning of Sergei Skripal or Alexei Navalny with a chemical substance.

For disinformation activities, Russian intelligence use internet platforms, conventional media, agents of influence, and "useful idiots" – individuals who unknowingly spread disinformation messages. Their task is to disseminate clear but deceptive messages proving the defensive nature of Russian actions opposing aggression from the U.S., NATO, and Western countries. It is the base narrative for a multitude of stories with variable plots, dependent on context, place of circulation, and target audience. Within this framework, there are "little lies" denoting the distortion of facts or their biased interpretations. However, there are also "big lies", meaning the pure creation of facts on an unimaginably large scale. This is done to replace the reading of reality by recreating it. A classic example of such behavior is the Kremlin's disinformation surrounding the mobilization of troops on the Russian-Ukrainian border at the turn of 2021 and 2022. It focused on justifying this fact by citing Ukrainian aggression materially supported by the U.S. and the West, and the offensive deployment of troops and weapons by NATO on its eastern flank. These lies then entered the spaces of media and scientific and political communication in democratic countries, effectively widening the existing social divisions by exploiting gaps in knowledge and the inability to verify facts.

Russian disinformation campaigns currently pose the main threat to Poland's security in the information sphere. Since 2014, Polish security services have observed a significant increase in offensive activities in cyberspace by entities associated with Russian state structures or those acting on their behalf. This includes mass information and psychological operations aimed at NATO and EU countries. Since 2020, these activities have been largely supported by Russia's ally, Belarus, whose authorities initiated a disinformation campaign against Poland, Lithuania, Ukraine, and the United States after Western states refused to recognize the presidential election results falsified by Aleksandr Lukashenko (Bryjka, 2021a, pp. 157–180). Disinformation activities intensified during the border crisis caused by the Belarusian authorities at the turn of 2021 and 2022 when the weaponization of migration was used as a tool of pressure on NATO and EU countries (Bryjka & Legucka, 2021a).

A 2014 report by the Polish Internal Security Agency emphasized the increased activity of Russian intelligence services conducting disinformation activities subordinated to the Kremlin propaganda strategy that were conditioned to the conflict in Ukraine. Their goal was to discredit the position of Poland and other NATO member states regarding the Ukrainian crisis and emphasize complex historical experiences between Poland and Ukraine to provoke antagonisms between the societies of both countries.

Another element of Russian activity seeks to highlight and create divisions among NATO and EU countries. To implement such projects, Russia has used controlled media and Polish citizens who represent a pro-Russian attitude (*Raport*, 2015). A similar situation has been assessed by the counterintelligence services of other countries in the region, including Lithuania, Latvia, Estonia, and Czechia (Baraniuk, 2017).

In the report for 2015–2019, the Internal Security Agency emphasized that Poland is particularly exposed to hybrid activities carried out in the information space with the use of social media. As the basic tools of information warfare, they are aimed at Polish security and the state's image. Their goal is to distort the political and social situation and exploit extremism to introduce divisions among citizens and polarize public opinion. Between 2015–2019, Poland expelled 28 foreigners for hybrid-type activities against national security (*Ochrona*, 2020). In March 2021, for example, at the request of the Military Counterintelligence Service, a Russian citizen claiming to be TV reporter Yevgeny Reszetnev was recognized as *persona non grata*. According to Polish authorities, his journalistic profession served only as a cover for espionage activities. Materials obtained by Polish military counterintelligence officers proved that the Russian agent's task was to collect information for use in disinformation operations against Poland and NATO. At the end of February 2022, the Polish Internal Security Agency also detained a Spanish citizen of Russian origin working for the GRU. The accused pretended to be a journalist in Warsaw and at the border with Ukraine. His task was to collect information that could have a negative impact on Polish security and defense efforts and to spread disinformation activities concerning refugees from Ukraine to distort the image of the situation on the border.

Russian disinformation operations like those against Poland are part of the Kremlin's subversive policy towards NATO and EU countries, which aims to weaken the transatlantic community. Poland is especially relevant in the Russian strategy of disintegration in the West due to its geographic location, historical experience, different value system, and conflicting interests in Eastern Europe. Disinformation is therefore a direct and indirect threat to the national security of the Republic of Poland as it has a destabilizing effect on the internal situation in Poland and the international environment in which Poland functions. The militarization of information is deeply rooted in Russian strategic culture, and there are no indications that hostile actions against Poland in the information sphere will be limited.

The current dynamic security situation in Europe points to a likely intensification of disinformation campaigns against Poland not only by Russia but also by Belarus, which is imitating Russia and China by operating in the information space at an increasing scale. Due to these conditions, the demand for analysts of disinformation is growing. State structures, in cooperation with the private sector and the expert and scientific communities, needs to cooperate to build a system of state resistance to disinformation.

Counteracting information manipulation cannot be limited to actions undertaken by state structures. It should use the wider potential of society, giving it specific tools and opportunities for a bottom-up approach of building counter-disinformation capabilities. This could be done, for instance, by providing grants for projects that monitor disinformation from various directions (e.g., Russia, China, and Belarus); financing research in the field of new technologies that allow for the detection and combating of disinformation through artificial intelligence; offering financial support for training analysts; and integrating other sectors to involve them in public administration initiatives.

Summary

1. What constitutes state information security?

The state's information security includes: (1) the protection of its information resources, especially classified information; (2) the protection of state institutions and society from the influence of disinformation and propaganda; and (3) developing offensive abilities against the information resources of a potential adversary that could have an influence on society.

2. What distinguishes agents of influence from useful idiots, trolls, and bots?

Agents of influence are people who duplicate disinformation on behalf of a foreign intelligence service. They can receive certain benefits, which are often financial, for their activities. "Useful idiots" also propagate foreign disinformation and propaganda but do so unconsciously. Their actions are not the result of a task or motivation from a foreign intelligence officer but are guided by their personal views, beliefs, or carelessness. Trolls are people who shape the debate on the internet by publishing comments and entries on behalf of state or private entities. They receive compensation for their activities, and their work is controlled and coordinated by their superiors. Bots perform the same tasks as trolls but are created through artificial intelligence algorithms that automatically post entries and comments.

3. What are the main types of information falsehood?

"Misinformation" is information that does not correspond to reality. It can be disseminated both intentionally and unintentionally. "Disinformation" is information that is deliberately created or reproduced and which is untrue or manipulated and used to mislead the recipient for specific political, economic, or military purposes. "Malinformation" is the misuse of information used, for example, to stigmatize certain social groups through hate speech.

4. What is the Russian scheme of interference in elections in democratic countries?

The Russian scheme of interference in the political processes of democratic states includes cyber-attacks on information systems of political parties and government structures; hacking of private and business emails followed by controlled leaks of stolen data (hack-and-leak); and mass dissemination of disinformation on social media using bots, trolls, propaganda tubes, networks of pro-Russian think-tanks, agents of influence, and other active measures. The

 | DOI:10.1163/9789004715769_012

aim of these activities is to exacerbate existing social tensions, undermine citizens' trust in democratic institutions, and promote people and political groups that are friendly or neutral toward Russia. It does so to discredit opponents and create an atmosphere of chaos and uncertainty.

5. What are the basic principles of disinformation in the RESIST model?

The basic principles of disinformation can be described by the acronym FIRST, which stands for: Fabrication and manipulation of the content of the message; identity, which involves concealing or stealing an identity; rhetoric, or using arguments based on false information or offensive attacks; symbolism, which involves the use of symbols to enhance the communication message; and technology, which means taking advantage of new technologies like bots.

6. What are the methods and techniques of the Russian disinformation model (SWAMPED) identified by the EUvsDisinfo team?

The EUvsDisinfo team has identified the following instruments of Russian disinformation, referred to as SWAMPED: Strawman, or attacking opponents for views or ideas they never expressed; whataboutism, which is a technique of responding to an accusation or difficult question by making a counter-accusation or raising another issue; attack, which involves using offensive language to discourage opposition; mockery, or the use of sarcasm to undermine an opponent; provocations, which uses charges that are both provocative and baseless; exhaust, which involves flooding an opponent with details and technicalities; and denial, or questioning any evidence of the existence of disinformation.

7. What is critical thinking?

Critical thinking is the reflection and analysis used to make decisions and solve problems. It is based on logical, careful inference and is a process driven by good evidence. It defines problems, identifies conflicting arguments, uses relevant and reliable data, raises fundamental questions, and uses information efficiently to make judgments and informed decisions. Critical thinking is directly aimed at achieving a well-founded opinion and using adequate standards of evaluation that determine the true meaning or value of something.

8. What principles should fact-checking organizations follow?

The basic principles are contained in a code of fact-checking organizations established by the International Fact-Checking Network (IFCN). Their foundations consist of three elements: impartiality and honesty; operational transparency of sources, funding, organization and methodology; and an intention to improve the quality of public debate.

9. What elements does the simplified disinformation recognition model consist of?

The International Federation of Library Associations and Institutions (IFLA) has proposed a simplified model for identifying false information consisting of the following steps: (1) checking the source; (2) checking the authors; (3) checking the date of publication; (4) reading the entire text; (5) paying attention to bias; (6) verifying with additional sources; (7) checking for satire; and (8) confronting the news with the opinions of experts.

10. What are basic principles of operational security (OPSEC) in open-source intelligence (OSINT)?

Basic OPSEC best practices include: (1) using a password manager and two-factor authentication; (2) using anti-virus software; (3) systematically updating operating systems, applications and programs; (4) systematically backing up research to an external drive or in the cloud; (5) selective sharing of information about oneself on the web and limiting one's visibility to others on social media; (6) protecting sensitive information concerning one's own identity and activities; (7) reducing digital traces left in the network; (8) avoiding the use of open Wi-Fi networks; (9) using encryption keys and secure applications; and (10) anonymizing activities through the use of a VPN or TOR browser.

PART 3

Disinformation: Countering and Resilience

CHAPTER 9

Countering Disinformation: General Characteristics and Immunity Building

In Parts 1 and 2, we characterized disinformation as a contemporary threat to international security and presented the conditions and tools for its identification. In Part 3, we strive to answer crucial questions related to countering disinformation in international politics. First, we address whether we can combat it effectively in the modern online world, and if so, who should be responsible for doing so. Next, we examine whether the general consensus is that disinformation is harmful. As part of this question, we also examine what it means to build resilience to disinformation and we evaluate its importance at the individual, national, and global levels. Finally, we explain how disinformation is linked in the national and international dimensions.

The age of the internet has brought about a weaponization of disinformation, which is now being utilized by both state and non-state actors in international politics due to its high speed, low cost, and effectiveness in hybrid operations. Disinformation harms individuals, social groups, states, and international organizations, and its perpetrators exploit the global reach and speed of social media and online platforms. Among other Western democratic states, Poland is at the forefront of such actions, particularly against Moscow. Russia's propaganda aggression against Poland has intensified in parallel with its armed aggression against Ukraine. This aggression against the Polish state image poses a threat to national security by undermining its credibility in NATO and the EU, as well as its position in the Western community and the values of democracy (Kupiecki et al., 2021). Poland, its authorities, and Polish society must take a suitable approach that includes both prevention measures and response. Membership in the European Union and NATO supports Poland's efforts in defending against Russian propaganda. This membership means that Poland can make the most of working together with its allies and partners and that it can use good practices and mechanisms of policy coordination to strengthen public resilience to disinformation. Ultimately, however, each country is responsible for its own resilience and response measures.

DOI:10.1163/9789004715769_013

1 Fighting Disinformation: General Characteristics

The way that foreign actors now use disinformation is reminiscent of older methods, such as traditional propaganda, lies, corruption, and recruiting agents for disinformation purposes. Past measures and operational undertakings by secret services that counterintelligence once faced pale in comparison to today's atmosphere. Secret services still keep their arsenal stocked with several hundred thousand leaflets, brochures and forged documents, in combination with using methods such as infiltrating specific milieus and recruiting spies. However, the internet has made it possible to influence the activities of millions of people quickly, cheaply, massively, and effectively without the use of old and outdated methods (Rid, 2020). Moreover, people's unconscious complicity in creating their own messages or duplicating foreign ones further contributes to this influence. Depending on the needs of the information environment, fake accounts as well as active and dormant bots are employed to achieve this goal. While traditional media environments and disinformation tactics are still used, the internet has exponentially increased the field of information warfare.

Social networks are often blamed for the prevalence of disinformation among internet users. Documents and letters written or signed by politicians or military members are being forged and distributed online. This new digital reality is rapidly transforming the world. Renée DiResta of Stanford University has coined the term "ampliganda," derived from amplified propaganda, to describe this new environment. She contrasts it with disinformation as a broader and more dangerous phenomenon that can cause avalanche effects through the use of social media, clickbait, and hashtags, even without the user's ill will or awareness of duplicating false content (DiResta, 2021).

Traditional knowledge – of the enemy's goals, methods, techniques, codes of intelligence, and diplomatic and political culture – is not enough to combat such amplified disinformation. New abilities, skills, and resources are required and, given the nature of the threat, security services or state institutions alone cannot be responsible for counterattacks. A holistic and integrated approach is essential to tackle disinformation of such magnitude. This approach involves not only the mobilization of state structures, the appointment of inter-ministerial teams, and the adoption of government counteraction strategies and programs but also the participation of civil society, social organizations, researchers, scientists, journalists, teachers, and ordinary internet users. It requires a synergistic combination of these resources in national and international counteraction, with digital civil defense as one of its main components. The importance of such spontaneous actions on a global scale can be seen in the international community's response to the Russian invasion of Ukraine. The response included a mass campaign to inform Russians through text messages, emails, and telephone calls about actual events on the front line. It also included a grassroots action by international hackers who coordinated their efforts to block the websites of Russian institutions and their propaganda networks.

Effective defense in the digital world must begin with an understanding of the mechanisms of how false news are spread. State services and expert circles must be familiar with effective methods of countering disinformation. However, to a great extent this defense also depends on the attitudes of individual participants in the infosphere, their digital education, and their understanding of the contemporary conditions of their own cognitive security. Ideally, everyone should understand not only why some people lie in the public space but also why and for what purpose lies are spread *en masse* and with no serious control. This is the first step toward becoming immune to falsehoods and developing critical and effective media literacy. A basic knowledge of psychology, computer science, and pedagogy is valuable in building the skills to navigate the information space. However, it is unreasonable to expect people to gain this understanding on their own. Media literacy should therefore be included in school education from an early age and in all public education. An increasing number of countries are recognizing the importance of media education and are integrating it into their educational programs. Finland is considered a model country in this regard as it implements educational solutions from pre-school to higher education. The Finnish model works because young children are exposed to the internet from a young age.

The need for media education for young people is further confirmed by an analysis done by the American RAND Corporation. The findings included over 200 research reports, 64 of which were in-depth, from around the world. The reports were reviewed to identify recommendations on how to counter disinformation. One of the obvious conclusions was the need to develop media education programs that teach people to use media wisely, especially in the world of social media. Such programs are crucial for educating all generations (Helmus & Keep, 2021).

Decision-makers responsible for implementing programs to combat disinformation should not be discouraged by skeptical expert opinions. Some experts claim that information manipulation does not significantly impact social behavior or strategic decisions in foreign and security policy, as mentioned in the first part of this book. If this were true, however, it would demonstrate the massive immunity of democratic systems. The lack of effect of disinformation operations would discourage their organizers from taking similar actions in the future. It would also demystify the risks associated with them and alleviate concerns about the credibility of electoral processes, the legitimacy of elected authorities, and the security and political stability of states or the state of democracy in the Western community. Unfortunately, the current reality does not support views that disinformation is ineffective (Bateman et al., 2021; Posard et al., 2021).

In discussions about the ineffectiveness of external disinformation, however, an important theme is the responsibility of Western societies for this phenomenon. Their responsibility is supposed to result from internal social divisions exacerbated by social media. It is difficult to argue with this thesis, even if it is not easy to measure the cause-and-effect dependence of social polarization on digital interference from abroad. Consideration of independent studies and government reports in states affected by informational aggression reveals that the scale of this phenomenon is currently unprecedented in international relations and poses actual and potential threats to democracy worldwide, regardless of the internal conditions of societies (McInnis & Starling, 2021).

Assessing the effects of influence operations and countering them will always be difficult. In general, however, the social and political consequences of disinformation are clear; for example, its effect on democratic elections. Disinformation has led to several related problems, including the deepening of divisions and differences of opinion, internal tensions, disinterest in participating in the election process, and reduced confidence in public institutions. Even established democracies, like the United States during the presidency of Donald Trump or the United Kingdom during the Brexit referendum campaign, have been affected by disinformation. Canadian special service reports have also identified constant targets for disinformation (Carvin, 2021). Similarly, analyses of foreign interference preceding the parliamentary elections in Germany in September 2021 highlighted similar risks. Disinformation was one of the key instruments used by Russian attackers on German democratic institutions. Fortunately, Germany thwarted the disinformation attempts thanks to thorough preparation and responses from the part of state structures and civil society.

During the 2016 U.S. election campaign, pro-Kremlin Facebook accounts organized demonstrations by two opposing groups, pro- and anti-immigrant, to incite riots as part of the Heart of Texas operation. Fortunately, casualties were avoided, but the goal of increasing local tensions and disputes was achieved. The outcome was also used to criticize American democracy and diminish its image abroad. More tragic consequences followed the presidential election in November 2020. Street protests grew at an unprecedented pace, fueled by emotional comments on social media by Trump supporters reacting to the election results. Within 24 hours, the Facebook group *Stop the Steal* mobilized over 300,000 people and activists convinced that Joe Biden's supporters rigged the presidential election or, in their words, "stole the presidency". While we do not know the extent to which foreign countries and activists acting on their behalf were involved in this operation, history teaches us that they did not remain idle (Sweet, 2021). In part 2 of this book, we discuss the political and strategic goals of disinformation, regardless of its strictly military dimension, in the context of Russia's attack on Ukraine.

With the social polarization, tensions, and associated strong emotions that come with online networks, mobilizing people through social media is simple and effective. Fast and extreme messages resonate with users who are already awakened to a particular issue. People leave many traces of everyday activity on social media that testify to their political or election sympathies. Mass data sets are processed using software that allows for defining the preferences of users and adjusting the messages, whether commercial or political, that appear on their screens. Algorithms evaluate, organize, and provide detailed information about users and their preferences. They decide what information and content will engage recipients the most and encourage them to click and like content, which in turn generates viewership, economic profits, or election victories. Recipients become individual addressees, pushed to buy a specific product, express a desired opinion, or support a candidate or political party.

To change the functionality of social media used by manipulators, we must promote content capable of engaging various sides of the political and social spectrum in a civilized discourse. This will prevent manipulators from being rewarded with visibility on the web through posts that manipulate emotions or target basic needs. Essentially, platforms such as Facebook should be less provocative and more conducive to public debate. With less hate speech and aggression on the internet, the field of activity for foreign manipulators will automatically narrow. However, the social good does not necessarily align with the business goals of internet platforms. It is not altruism that drives these platforms but rather the financial measurability of actions. In the absence of top-down regulations and restrictions, financial incentives are their main motivation.

The fundamental premise is that, as part of a comprehensive approach to the problem, online platforms, especially large social networking sites such as Facebook, Twitter, or YouTube, should be obligated to counteract emotions on the web. In other words, technological solutions limiting the spread of hate, misinformation, and extremism should be utilized. The problem lies in the scale and effectiveness of such activities. For example, YouTube deletes approximately 10 million videos per quarter, but this is only an approximation as YouTube does not provide full data. In the first dozen or so months of the COVID-19 pandemic, over a million posts related to the coronavirus were removed. Interference in user-generated content is not smooth sailing for moderators because it can be difficult to distinguish between fake news and beliefs. This exacerbates the problem of equating truth and falsehood, but it also raises the question of how to make distinctions when the case is not obvious.

Nevertheless, the social media industry cannot avoid co-responsibility for the effects of information manipulation as they are often a component of influence operations that involve various other instruments of political influence. These include cyber infrastructure attacks, economic and energy blackmail, political corruption, espionage activities, and military measures that may lead to open military aggression. The January 2022 report of the special commission of the European Parliament investigating foreign interference in the internal affairs of EU Member States highlighted these aspects (*EU should,* 2022).

External interference in the internal affairs of states may not necessarily aim to impact electoral processes but to deepen or induce social polarization, impose or strengthen specific narratives, exacerbate public debate, or blur the line between facts and opinions. Such interference can also be part of preparations for the future. One example is the operation carried out in Ukraine in 2014 by *CyberBerkut*, most likely associated with GRU, which aimed to influence voters and make them doubt the process and the results.

In response to this phenomenon, the European Union and NATO have taken more robust actions against disinformation, with the primary aim of strengthening the social resilience of their member states and partners. These decisions were made quickly and without hesitation following Russia's military aggression against Ukraine, despite previous political and administrative barriers. Activities of state-owned Russian propaganda media, such as *RT* and *Sputnik*, have been banned within the European Union. Euro-Atlantic institutions view resilience to disinformation and other hybrid challenges as the ability not only to withstand and overcome challenges but also to do so in an irreversible, fair, and democratic manner. Such an approach places resilience in the context of fundamental values and goals, where problem-solving and crisis recovery do not occur at the cost of violating the guidelines of the rule of law and democracy, such as censorship of free media. Resilience, in this sense, begins with the awareness that if the authorities cannot defend themselves against such challenges, they also lose the trust of their citizens. This is particularly crucial for leaders to consider. However, even the best knowledge unsupported by action is insufficient.

To create an optimal system of responses to disinformation, both at national and international levels, multiple conditions must be fulfilled simultaneously. Governments, societies, and individuals require a set of instruments that are interconnected to build this system. When referring to foreign models at the national level, it is essential to consider that solutions that work in one country may not necessarily apply to another. For instance the Swedes, who do not have a large Russian minority, find it generally easier to detect and combat Russian disinformation than the Latvians, given that a Russian minority constitutes a quarter of the entire population and half of the population of the capital city, Riga.

Regardless of different national approaches, it is important to recognize that the key link to resilience against disinformation is the individual recipient of these messages, who can act as the first line of defense in the information struggle. It is crucial for internet users, especially public officials, to understand why disinformation exists, how to avoid spreading it, and how to use media, particularly online. They should also be aware of the methods used by disinformers and practice basic *digital hygiene* to reduce the risk of intrusion into their communication tools. Media education and training can also improve individuals' ability to verify facts, and fact-checking is becoming increasingly accessible not just to experts or journalists but to ordinary internet users as well. By engaging in individual or group crowdsourced fact-checking, individuals can more effectively combat conspiracy theories and disinformation (Corlin & Önnerfors, 2021).

The harmful nature of disinformation has become increasingly apparent on a global scale due to the COVID-19 pandemic and hostilities aimed at Ukraine. False or manipulated information about the rate of the disease spreading and new mutations of the virus in India led to fear in countries thousands of miles away. Anti-vaccine protests were organized in Germany using social networks in Australia, and anti-vaccine misinformation spread around the world, especially on TikTok among teenagers (Fox, 2021). The popularity of new social media sites, such as Rumble and Gap (Kaplan, 2021), also increased. The pandemic highlighted the unprecedented global scale of deliberate manipulation of information by some states and non-state actors and resulted in a greater awareness of the impact of falsehoods on the web and other forms of communication. As a result, national and international actors have taken more vigorous actions, leading some platforms to consent to the voluntary EU Code on Counteracting Disinformation.

During the pandemic, conspiracy theories and xenophobic beliefs gained momentum (Hellerstein, 2021a). The Russian FSB certainly does not pay all anti-vaccine workers recruited from medical circles in Great Britain, Germany, Poland, or other countries. It is no coincidence, however, that many activists in pro-Kremlin circles or internet influencers favoring the FSB have become active opponents of vaccination while also promoting conspiracy theories and anti-EU slogans. This convergence served two purposes. Firstly, it enabled them to deepen social divides and erode public trust in institutions through the use of a popular topic that evoked strong, extreme emotions. Secondly, they broadened their sphere of influence by taking advantage of numerous "useful idiots" who put themselves in such a position. The essence of this phenomenon is not the level of their association with an aggressive state or non-state actor but rather the objective convergence of propagated and socially harmful content. Disinformers paid by external entities and spontaneous disinformers can cause society similar harm. Due to their minimalist moderation policies and ultra-liberal regulations, encrypted platforms such as Telegram have already attracted oppositionists, conspirators, and liars. Telegram in Germany became a hotbed of false information about COVID-19, as well as racist and anti-Semitic content during the pandemic. As of the end of 2021, the platform had around 1.5 million users in Germany, Austria, and Switzerland (Wildon & Gildejeva, 2021). Since July 2020, Telegram has been used to coordinate approximately four thousand protests, which were primarily related to anti-vaccination sentiments. Russian broadcasters *RT* and *Sputnik*'s accounts have also grown more popular on the platform.

The methods and means of disinformation discussed in earlier parts of this book can be supplemented with the migration of internet users across social messaging platforms. After Twitter blocked former U.S. President Donald Trump's account, supporters flocked to the GETTR platform. He also planned to launch his own website to provide a place for outreach to his political and curious supporters. Despite claims that the movement is fading due to reduced social media presence, supporters of conspiracy theories and anti-establishment beliefs, such as QAnon, have not disappeared; instead, they have migrated to other sites, including those involving foreign influence agents. Changes in the tools employed can also be observed in Estonia and Ukraine, where pro-Kremlin media lost their licenses and attempted to resume activity under new names. Following the failure of the local version of *Sputnik* in Sweden due to little interest from recipients, Russian or affiliated entities purchased local online media to domesticate operations of influence, disseminating foreign content and narratives through seemingly native sources. Russian and other disinformation organizers are increasingly utilizing the growing market for commercial services in this "industry", both in target countries for information operations and in countries where service providers may operate with a wider geographic scope.

Changes in disinformation tactics and operational methods are linked to advancements in data processing capabilities, creating the possibility of even faster, more insidious, and harder-to-detect distortions of reality. For instance, a detected network of 350 fake accounts spreading pro-China narratives in the French news community featured profiles created using artificial intelligence (Carmichael, 2021), which highlights the new challenges and threats emerging. Deepfake technology, which allows for the creation of fake images or videos using artificial intelligence, presents ethical dilemmas in cases like the one of a deceased artist whose voice was used in a documentary about him. The manipulation of recordings featuring politicians and public figures is politically dangerous, especially when deepfakes aim to ridicule, discredit, or incite street protests and violence.

The combination of disinformation and technology creates new cyber threats that are increasingly complex and manipulative, targeting national security and election processes. In 2016, hacker attacks targeted the Democratic Election Committee and John Podesta, Hillary Clinton's campaign manager, during the U.S. election campaign. In 2021, the Russian military services GRU-associated group, APT29 (also known as *Cozy Bear*), launched an attack on the Republican Party. They did this simultaneously with cyber-attacks by criminal hacker organizations on critical infrastructure, the DarkSide group and the oil Colonial Pipeline, and the business entities, REvill and Kaseya (Turton & Jacobs, 2021). These events have raised concerns about the security of critical systems and led to criticisms of authorities for ineffective countermeasures. Despite attempts to resolve these problems through dialogue, further attacks continue. Russia's asymmetrical actions aim to balance the potentials of both countries, as illustrated by Putin's plan to create a national cyber-wall modeled after China, which would give the government an information monopoly and censor citizens from external content. The Kremlin has already censored content related to Russian military operations in Ukraine and introduced penalties for activities deemed improper or anti-state, including the use of the words "war of aggression" in online communication.

In the West, on the other hand, experience is being accumulated, studies and analytical volumes are increasing, and as a result, there is a better overall understanding of the disinformation environment and its actors, goals, and methods. In addition, there is a growing number of social and journalistic initiatives focused on media education and fact-checking. While the group of global disinformers is growing in the aftermath of the COVID-19 pandemic and Russia's attack on Ukraine, there is also a growing number of organizations and individuals involved in fighting disinformation. Public opinion in European Union countries appears to be assigning the related tasks equally among those in power, the media, and civil society, which is a good starting point for coordinating state actions and societal initiatives.

2 The Challenges of Regulating Online Platforms

Growing awareness of the risks and pressure from experts and societies are leading to new political and legislative initiatives on both sides of the Atlantic, with a view to regulating social media and online platforms more consistently than ever before. They entail:

- Regulation of the social media business model.
- Incorporating transparent policies and practices.
- Protecting users' personal data.
- Implementing transparency practices around political advertising mechanisms on the web.
- Programming algorithms that regulate the way platforms function in interaction with and between users.

At the same time, there is an ongoing discussion regarding possible agreement (and international coordination) between state regulators and the owners of global media and internet platforms. In the first decade of this century, the U.S. was of the option that the solution to misinformation was providing more information rather than limiting its spread and filtering out harmful content. This viewpoint shifted during the pandemic, however, when YouTube, Facebook, and Twitter started moderating the content on their platforms more rigorously. In response to countering Russian military propaganda against Ukraine, further measures were taken, which will be discussed later in this book.

According to analysts, global internet corporations like Facebook face a complex problem that extends beyond the technology of their media platforms. With 3.5 billion global users, Facebook is considered to be the largest information autarkic entity in the world. It collects various data from each user, processes it, and uses it as commercial and informational products. Mark Zuckerberg, the CEO of Facebook, has compared the potential of his company to that of a superpower and has even boasted about his company's rivalry with China in the production of cryptocurrency. However, the issue of power and control over privacy, identity, and information processes has become increasingly apparent. As a result, Facebook appointing teams such as supervisory boards or electoral issue commissions to solve the problems of hate speech and disinformation actually conceals a much more serious problem.

These actions are not enough, but they are a step in the right direction. At the same time, we must not lose sight of the fact that time works in favor of increasingly powerful corporations at the expense of global social interest. Critics of this model accuse Facebook of "ostentatiously abandoning morality" and using lies, predicting the collapse of our current civilization as a result. Facebook has failed to adequately respond to hate, propaganda, conspiratorial visions of history, and even terrorist recruitment carried out on user profiles. The users of the platform who are media-educated and aware of the risks could provide the greatest potential counterweight to such practices. Nevertheless, even if half of users were to abandon Facebook, about a quarter of humanity would remain within the circle of influence of Facebook's authorities (LaFrance, 2021).

With the complexities of these issues, changes in the functions of social media present a challenge for Western countries. Additionally, there is a concern that regulating online platforms could lead to their demise. Controlling the negative aspects of their activities could risk destroying the freedoms they provide. There are also concerns about how to censor undesirable content while not undermining their position in undemocratic countries where they serve as a source of alternative information for state propaganda, and where they face pressure and sometimes blackmail from governments. The dilemma of promoting freedom or preserving profits remains another concern.

As the example of Russia shows, the differences in the ways of managing digital space are deepening between democratic countries that aim to protect the rights and freedoms of individuals and authoritarian countries that prioritize the interests of the authorities. It is currently still unclear where internet segmentation may lead and how it will affect freedom and democracy, as it could create closed or disintegrated information spaces on the web. However, the problem of closed information spaces clearly weakens the role of the internet as a tool for promoting democratic values not only in China but also in Russia and many other smaller dictatorial or authoritarian countries.

3 Humankind in the Face of the Threat of Disinformation

In such a complex information, economic, and political environment, the individual level remains key in building resilience to disinformation. This is where the problem of the impact of disinformation and the weaknesses of prevention come into play. It is important to remember the reasons for telling untruths discussed in previous chapters. Researchers of the subject claim that being dishonest is so deeply ingrained in human nature that it may tempt one to say, "I lie, therefore I am". This statement does not have to be less true than the Cartesian "I think, therefore I am", and perhaps it is even a paradoxical complement to it. Lying requires thinking, and disinformation requires significant knowledge about people and society. At the same time, this thinking is also the first line of defense against disinformation.

Psychologists suggest that the ability to manipulate without the use of physical force has likely been advantageous in the competition for resources and partners, much like the evolution of deceptive strategies in the animal kingdom, such as camouflage. Sissela Bok, an ethicist at Harvard University, has noted that "Lying is so simple compared to any other means of gaining power" (Fullinwieder, 2007). In essence, there is no need for weapons when lying can be easily employed. Disinformation crafted and disseminated by a single person, acting under state control, can reach millions and achieve the desired effect. When choosing his career, Vladimir Putin appeared to be fascinated by this peculiarly understood minimalism, in which one person's influence can have a significant impact on the course of history.

Analysts have identified that misinformation is often perpetuated by mental shortcuts, cognitive biases, and illusions that discourage critical thinking and fact-checking. This has led to growing concerns about the harmful effects of disinformation and propaganda online. To address this challenge, tech giants like Google have partnered with research institutions to develop innovative approaches to combatting extremism and disinformation. For example, Jigsaw, a research unit of Google, has collaborated with the Laboratory for Research on Polarization and Extremism at American University to test the effectiveness of psychological inoculation. This is also known as attitude inoculation which is a technique that helps people to resist manipulated messages (Courchesne et al., 2021). A study was conducted to specifically test the effectiveness of psychological inoculation against narratives of male supremacy and white supremacy. The experiment aimed to build psychological immunity by exposing subjects to a weakened or persuasive message and causing them to reject it. This controversial method was therefore about the use of some information manipulation to build resistance to more harmful forms.

Eight hundred participants, mostly white men aged 18 to 35, were recruited for the study. This demographic group is often targeted by white supremacist propaganda. Participants were shown a short video containing rhetorical techniques characteristic of racist propaganda. They were then divided into five groups and presented with film messages of varying degrees of extremist narrative intensity: one with a blatantly racist message, another with a milder message, and a video without any racist content. The study showed promising results, indicating that people who watched the inoculation video were less likely to support scientific racism than those who were not in those groups. It also resulted in a lower level of willingness to provide ideological, financial, logistical, and armed support to extremists. Additionally, after watching the inoculation video, participants were more likely to be against extremist positions and perceived the sources of such propaganda as less credible.

Unfortunately, human psychology shows that the effects of disinformation can be lasting. In 2015, Australian researcher Briony Swire-Thompson conducted a study at the University of Western Australia which showed the ineffectiveness of providing evidence-based information in refuting false beliefs. The experiment involved presenting one of two statements to approximately 2,000 American adults: "Vaccines cause autism" and "Donald Trump said vaccines cause autism". Since Trump has repeatedly suggested that there is such a link despite lacking scientific evidence on the matter, it is unsurprising that Trump supporters showed greater confidence in the validity of this misinformation when his name was associated with it. After the initial test, the participants were presented with a brief explanation, citing a large-scale study, as to why the claim about a link between vaccines and autism is false. They were then asked for their opinion again, and this time, many accepted the explanation refuting the link. However, a week later, in subsequent tests, their belief in the misinformation had substantially reverted to its earlier level (Courchesne et al., 2021).

Other research (Courchesne et al., 2021) has shown that refuting lies with evidence can actually strengthen belief in those lies. Denying false information can sometimes be harmful and further the damage of disinformation. This is because people tend to believe that the information they have is true, or that they themselves have sufficient competence to verify it, as seen in the explanation of the Dunning-Kruger syndrome earlier. When they are challenged, there is a risk that the recipient will develop a psychological defensive reaction against a change of position. Paradoxically, this will make corrective information less effective in the long run.

The question that arises is how to correct erroneous opinions and beliefs without causing a yo-yo effect. To address this issue, a group of researchers from renowned research centers in various Western countries created a manual with a positive message. The manual assumes that, while it is better to act preventively as part of appropriate communication strategies in the long term, denying fake news can be effective when following a few clear rules: clearly state the facts, explain disinformation without scientific or professional jargon, and ensure that the denial is provided by competent individuals (*Debunking Handbook*, 2020).

Communication strategies and media education for individuals will undoubtedly aid in combating the new realities of the information environment saturated with emotions and manipulated information. The issues with digital platforms and social media are not new in terms of friction, tensions, trade-offs between freedom of expression and regulation, copyright, data protection, and accessibility. However, due to the nature of technology and digital space, these platforms create specific challenges related to the spread of disinformation that are unlike other media. In their information ecosystem, there is a toxic combination of four phenomena:

– Aggregation of data about each user and their behavior.
– Data algorithmizing, or data management with the use of computer programs with special processing capacity.
– Anonymity of aggregation, management, and dissemination of information.
– Automation of content publication and interaction with it and other users (Bradshaw, 2020).

This combination of technology has enabled both the public dissemination of disinformation and the corporations profiting from it. With an annual revenue of $100 billion USD, Meta's (the owner of Facebook) income surpasses the national product of most countries in the world. Its long-term influence, beyond its financial power, stems from the accessibility of data and the capability to employ it across all four of the above potentially harmful dimensions.

In data transmission, the traces of users' activity on the network allow platform managers to gather unregulated personal information, including data on users' political or election sympathies. Algorithms analyze this data and provide feedback to users based on their preferences, prompting them to engage with certain content. The profits of digital platforms from personalized and more clickable content are constantly increasing. However, this approach limits the independence of internet users. Algorithms suggest content based on individual preferences, leading to decisions that increasingly move beyond the recipient, who is no longer a decision-maker but an object of socialization. The cognitive weakness of an individual makes them more susceptible to the effects of this mechanism.

Automation enables the use of fake accounts or bots programmed to engage much more frequently and consistently on platforms within set information parameters. In operations of influence, these bots act as generators of the popularity of content by interacting with and sharing it. They are also used to launch aggressive attacks on opponents with hate speech or manipulated information. The creators of these bots will benefit from technological advances, increasingly augmented by artificial intelligence. So far, the response from platforms, governments, and supranational regulators has focused on content moderation, involving third parties such as journalists, media groups, researchers, and civil society experts at low costs through outsourcing. However, this model seems unsustainable in the long run without addressing the technological and systemic causes and methods of falsehood proliferation on the web.

Such changes are imperative because societies generally place their trust in internet media regardless of their awareness of the potential for manipulation. Research from the past indicates that people who received information through press, radio, and television, including official messages, also trusted these sources and considered them credible. However, with the current speed of internet broadcasts, the proliferation of sources, and the de-monopolization of the roles of official broadcasters, people have remained uncritically trusting of sources, particularly among the older generation. In other words, the vast majority of internet users believe in the information distributed, which they then willingly and naturally pass on.

Research on the effectiveness of platforms' responses to disinformation and influence operations is scarce and often inadequately documented. Some general conclusions can be drawn based on studies that employ sound methodology and reliable research questions that allow for statistically reliable comparisons of results related to the real world of international politics (Bateman et al., 2021). The most significant findings highlight the importance of fact-checking, warnings against disinformation, and the provision of credible sources. However, it should be acknowledged that knowledge about the effectiveness of other methods used to counteract information manipulation is still relatively limited. This applies to measures such as:

- Deterring and disrupting manipulation.
- Strengthening the moderation of network activity.
- Changing algorithms.
- Limiting microtargeting.
- Building social credibility through strengthening of the media.
- Changing the incentive system by developing antitrust activities and using de-monetization.

The research concerned mostly American internet users and was largely focused on Facebook and Twitter, leaving the matter open for the research of other platforms along with their geographic, cultural, linguistic, and thematic dilemmas (Bateman, 2021).

Due to the natural cognitive limitations of humans and the challenges arising from the business models of online platforms, social media users encounter numerous obstacles in their individual struggle against disinformation. State and societal action is therefore necessary, which could be based on existing models of success. Examples of the most effective approaches are found in countries that engage potential users and promote their self-organization into social and professional groups.

Civil society plays a key role in shaping the desired information space. It can offer innovative, practical solutions in specific social and cultural contexts. Its activities raise awareness and civic engagement both online and offline and increase the competences of ordinary citizens to combat disinformation. Civil society also suffers from resource scarcity that industry, governments, and international organizations could address through increased support.

The international community has significant untapped synergetic potential, which will be discussed in more detail in Chapter 12. Due to the multitude of actors, goals, and methods of disinformation by domestic and foreign perpetrators, it is only through the joint action of the international community that effective solutions can be developed at both national and international levels. The objective is not to universalize solutions but to coordinate them with each other, jointly address general regulatory and technological problems, share best practices, and exchange information. Furthermore, the response to disinformation must be multifaceted and comprehensive, encompassing both institutions and people and resources. It should also be adapted to the local environment and challenges, which may differ in terms of the intensity of the problem and its local specificity.

To counter disinformation in international politics, it is crucial to gain a better understanding of its constituent features, including the goals, methods, and actors involved, and monitor their activities. International organizations, particularly those in the Western world, play a crucial role in this effort. NATO and the European Union rely on studying and understanding the information environment, including its participants, narratives, message content, and impact on their member states. They evaluate these factors within the context of threats to their member states and use this information as their first line of defense against disinformation.

In the aftermath of Russia's aggression against Ukraine in 2014, NATO's communication strategy took on a new dimension, with countering disinformation becoming an even higher priority in the wake of the 2022 invasion. NATO has been developing an analytical project for many years, using breakthrough technologies and the ability to evaluate large amounts of data to identify the behavior of both perpetrators and recipients of disinformation. This project has been successful in increasing the Alliance's ability to monitor its information environment, assess the situation, and take countermeasures with more precision.

Similarly, in the EU, the starting point was the groundbreaking decisions made in 2015 after the illegal annexation of Crimea and Russia's aggression against Ukraine. One of these decisions was the establishment of a specialized East StratCom task force in the European External Action Service (EEAS). The team's work involves identifying, analyzing, and creating a database of Russian disinformation, which in turn contributes to the development of an EU communication approach, policies, and strategies for countering disinformation. The EU diplomatic service has played a particularly important role in this effort. The European Union has since implemented several additional projects, including the European Action Plan for Democracy adopted in 2020. Furthermore, the legal act on digital services from 2022 was another milestone, imposing obligations on service providers regarding the transparency of users' activities, protection of their data, political campaigns on their profiles, advertisements, or content moderation on, in particular, the largest platforms such as Facebook, YouTube, and Twitter. The EU may also contribute to the promotion of the UN Convention on Universal Digital Human Rights in the future. The United Nations, its agencies, and other international organizations such as the Council of Europe, the OECD, and the OSCE also contribute to the fight against disinformation, though not as comprehensively as the EU.

The topics discussed in this chapter lead to the conclusion that countering disinformation is becoming a top priority in the policies of many countries and international organizations in the international arena. Moreover, there is a growing number of research and analytical centers, fact-checking organizations, and media education initiatives collaborating with each other in the non-governmental community. These entities also form important international clusters in regions and countries with less experience or response capacity. This environment will be discussed more comprehensively in the next chapter.

CHAPTER 10

Media Education and Counteracting Disinformation at the Individual Level

It is important to distinguish between media education and media literacy. Although these terms are often used interchangeably, they have different meanings. Media education goes beyond practical media literacy, and its primary objective should be to develop critical thinking skills. With the advent of social media, the ways in which we use information sources, build knowledge, and influence the information environment and its actors have been revolutionized. This is why the EU Audiovisual Media Services Directive (DAUM) has expanded the scope of media education and defined it as "skills, knowledge, and understanding that enable citizens to use media effectively and safely, think critically, evaluate, analyze complex realities, and differentiate opinions from facts" (*Directive of the European Parliament*, 2018).

1 The Role of Media Education in Countering Disinformation

There are no global, uniform standards and practices in media education, and there are many reasons for this. One reason is the varying levels of general education and the speed at which it adapts to changes in the environment, especially those related to desired skills. Another limitation is the lack of awareness about the importance of media education, as well as the systemic preparedness and conservatism of teachers. Many teachers are not always prepared to teach lessons as quickly as they receive new information due to the rapidly changing landscape of technology. Moreover, young internet users are quick to adapt, and it is not uncommon for students to have greater skills in informatics and using computers than their teachers in computer science lessons. This does not mean that students should assume the role of teacher, but it should prompt reflection on how, what, and why to teach.

In the U.S., there are several factors that contribute to inequality in media education, such as ethnicity, student wealth, location, school district, and school type. Moreover, American teachers have expressed concerns about the lack of guidance on modern teaching content. While approximately 41% of teachers incorporate media education elements into their lessons, only 17% treat media education as a separate subject. However, based on research,

DOI:10.1163/9789004715769_014

experience, and recommendations from individual American states, the implementation of new mandatory programs for media education and countering disinformation is on the rise. One of the trailblazers in this regard is the state of Illinois, which made media education obligatory (Baker et al., 2021).

The authors of NESET (McDougall et al., 2018), a comprehensive report prepared for the European Union titled *Teaching media literacy in Europe: evidence of effective school practices in primary and secondary education*, provides practical guidance on what and how to teach media literacy, what skills are necessary, how to measure them, and what education policies should be recommended throughout the EU. The ultimate goal is to guide students away from merely acquiring technical media skills towards developing pro-social and civic behaviors. According to the recommendations, the desired capabilities should include:

- Accessing, finding, using, and sharing information from a variety of sources.
- Understanding, critical evaluation, quality analysis, truthfulness, and credibility of views and effects.
- Creating content and expressing individuality in a way that inspires trust, with an awareness of the purpose and addressee.
- The ability to behave in a socially responsible and ethical manner, consistent with one's beliefs and intended communication goals, and the ability to consciously use the media.
- The ability of social, civic engagement through the media and political self-realization based on democratic values and attitudes.

The aim, therefore, is to develop a range of attitudes and skills that lead to pro-social and civic behaviors, covering the axis of recipient-creator of information-responsible citizens. To achieve this, the EU institutions are developing comprehensive programs that combine countering disinformation with media education and digital competencies. However, the NESET report highlights the challenges of combating disinformation without interfering with the basic beliefs of students. The report emphasizes the importance of preparing teachers and creating a competency framework, particularly in secondary schools. It suggests that correcting false information may be ineffective for students with strong convictions and recommends using multiple sources and viewpoints, fostering open dialogue, and avoiding hasty and definitive assessments.

Those responsible for school curricula and their implementation must decide whether to include media education as part of various subjects or as a separate subject that integrates content from different fields of knowledge. Media education can be included in social sciences, humanities, science, or natural sciences. It has not been uniformly practiced as a compulsory subject of education in EU Member States. In Finland, education on combating disinformation appears in many types of lessons. For instance, in history, students learn about distorting facts, events, and their interpretation; in mathematics, students learn about issues related to the falsification of statistics and figures; and in art classes, students observe examples of image manipulation.

Regardless of the model adopted, one approach does not have to exclude elements of the other. The recommendations of the authors of the report and other analyses propose developing media education programs: first, to develop media education programs that cover the five essential skills mentioned; and second, to methodologically support teachers and provide them with funds for implementing these programs. Media education should involve investment in pedagogical research, cooperation with civil society, journalists, and non-governmental organizations, as well as the inclusion of media education in teacher training and educational policies. It is also important to raise awareness of the dangers of disinformation and the risks associated with using internet content.

In the latter context, experts from the Polish National Broadcasting Council have made an interesting evaluation (*Fake news*, 2020). Referring to the need to develop media literacy, critical thinking, and protection of audiences from harmful content, they pointed to the obligations of providers set out in the aforementioned EU Directive on Audiovisual Media Services, amended in 2018, which entails first and foremost protecting their audience, including:

– Protecting underage users from inappropriate content in broadcasts, videos, and commercial audiovisual broadcasts by using and creating further appropriate provisions in the regulations. This may include marking commercial messages and inappropriate content, creating age verification systems for users, and implementing parental control systems.
– Protecting the public from user-created content and commercial messages inciting violence or hatred based on gender, race, color, ethnic or social origin, genetic characteristics, language, religion or belief, political opinion, belonging to a national minority, property, birth, disability, age, or sexual orientation.
– Protecting the public from content pertaining to public incitement to terrorism, child pornography, and racist and xenophobic crimes, the dissemination of which is a criminal offense under EU law.
– Creating and operating transparent and user-friendly mechanisms for reporting and assessing inappropriate content and informing the public about actions taken.

Second, the amended directive imposes obligations on the Member States: (1) periodic reporting to the European Commission with the first report by December 19, 2022, and then on a three-year basis after; and (2) establishing mechanisms to assess the suitability of the measures taken by video-sharing platform providers and entrust the national regulatory authority with the assessment of these measures.

Among the various national approaches to media education, some countries have model solutions, particularly in the Nordic countries such as Finland and Sweden. Lithuania, Latvia, and Estonia are also following in their footsteps. However, in most countries around the world, systemic media education either does not exist or is only just now being developed, particularly in regards to disinformation by external actors. It is easier to adapt policies and practical approaches to new challenges for those who have been teaching media use and disinformation in general for years. In Northern European countries, the approach to media education is based on a high level of general education, school autonomy, and integrating it into the curricula of the youngest and even pre-school students.

National approaches and practices in combating disinformation through media education can be classified according to various criteria, such as systematic and temporary, centralized and autonomous, and based on the presence of relevant content in the curricula. Regardless of the model, however, their effectiveness also correlates with the general wealth of societies and the level of education. This is indicated by research conducted by the Open Society Institute in Sofia (*The Medicine Literacy*, 2019) and as part of the International Student Skills Assessment Program (*21st-Century Readers*, 2021). Both studies also confirm that the effectiveness of media education is reduced in societies with higher levels of social exclusion, polarization, and tensions.

The aforementioned study by the Polish National Broadcasting Council provides examples that also refer to activities in education by national media regulators. It presents a narrow but interesting perspective on the nature of policies in this area.

In France, there is a systemic approach to media education, and a new impetus was given to it with the Act on Combating Information Manipulation of 2018. This act imposed an obligation to counteract the dissemination of false information on internet services. The French Audiovisual Regulatory Authority has the statutory right to issue recommendations to service providers, encouraging them to provide tools to identify credible sources of information, raise awareness of the impact of online content, and support the development of critical thinking, particularly in children and young people. The authority also recommends supporting projects and creating partnerships for media education, teaching information and digital education, creating independent initiatives by journalists and scientists to understand the phenomenon of disinformation, and conducting information campaigns. The

French education law encourages the development of technological education, critical thinking, information analysis, identifying threats related to social networks, and in the case of a higher education program in pedagogy, media education. France therefore serves as an example of an integrated approach to media education that is proactively implementing the recommendations in the EU directive at the national level.

In Ireland, which follows a civic model, media education is central to the mission of an organization that voluntarily brings together over 120 institutional members from various sectors for media education. The organization includes representatives from the Ombudsman's Office; the Film Institute; the Creative Europe Office; associations of librarians, advertisers, audiovisual, and media providers; and online platforms such as Facebook, Google, and Twitter.

In the United Kingdom, which also follows a systemic model, a program for the development of media education, called *Making Sense of Media*, was developed under the auspices of the national regulator, Ofcom. The purpose of the program and expert network is to promote online media literacy, research, and the coordination of activities in different areas of education. The expert network includes representatives from academia and non-governmental organizations. An example of such activity was the issuance of a report by Ofcom in cooperation with the UK Personal Data Protection Agency on *adtech technology* used in advertising strategies to target users. Since July 2019, Ofcom has been working with a team comprising representatives from UNESCO, the BBC, Facebook, Google, the National Association of Librarians, the Oxford Internet Institute, and the London School of Economics, among others. This team advises the regulator on matters related to new research areas and the evaluation of activities aimed at increasing media competences and critical thinking among citizens.

Media education is a part of the curriculum in most European Union countries, starting from elementary school level, and is mainly taught in subjects such as languages, history, social sciences, visual arts, social studies, and communication technologies. However, the extent and content of these classes can vary depending on the interests and preferences of the school or teacher, as is the case in Poland. Some countries, such as Czechia, Finland, and Hungary, have framework programs for media education at the national level, while the Netherlands, Germany, and the United Kingdom have adopted more systemic solutions, according to the NESET report of 2018 (McDougall et al., 2018).

Based on the changing needs of the environment and limited knowledge, the state of media education in the European Union can be assessed as moderately satisfactory, although teaching about disinformation by foreign states and non-state actors, particularly at higher levels of education, is limited. The prospects for improvement in this respect once again transpire from the decisions of the EU institutions. In the fall of 2021, the European Commission launched an expert group on disinformation and digital literacy to develop common teacher guidelines and curricula for citizens, as well as to create standardized education norms in Europe to combat disinformation and improve digital literacy. This initiative is part of the Commission's Digital Education Action Plan launched in 2021, and the group consists of 25 people, including researchers, teachers, representatives of NGOs, the media, and the cybersecurity sector. As part of the Digital Decade Roadmap, the Commission has set a target of ensuring that 80% of the EU population has basic digital skills by 2030, as opposed to the 56% documented in 2019 (Killeen, 2021a, 2021b).

Research conducted in the U.S. has identified significant gaps in media education and in particular, low levels of student competence and a lack of clear guidelines for teachers regarding curricula. RAND experts have prepared a special report on this matter, outlining the implementation and evaluation of a potential systemic educational program (Huguet et al., 2021). Within this framework, they distinguished six universal stages:

1. Defining learning expectations.
2. Identifying conditions that may affect educational efforts.
3. Exploring teaching potential, teacher availability, and resources.
4. Identifying teacher competence indicators.
5. Monitoring progress.
6. Measuring the overall impact of media education on a student's school performance.

This approach can be used to create integrated and outcome-verifiable media education programs in any country.

The state of media education in Poland appears to be inadequate in meeting current needs. By NESET and RAND standards, education authorities and teachers have yet to establish an effective media education program in the country. Furthermore, an informal survey conducted through interviews by the authors of this book with over 40 high school students, including recent graduates from 2020 and 2021, suggested that media education in Polish secondary schools is largely non-existent. There is a lack of appropriate school curricula, with even the outlines of such curricula being absent. The topic is only introduced if a teacher takes a personal interest and initiative or if students pursue it independently.

At the university level, several Polish universities offer classes on disinformation, including the University of Warsaw, Vistula University, and Collegium Civitas. In 2021, academics from the University of Warsaw, the Warsaw School of

Economics, and the Military University of Land Forces in Wroclaw launched the *Platform for Countering Disinformation* (PCD) *- Research and Education* with the support of NATO. The PCD aims to create an integrated project for countering disinformation at various levels of education and has published examples of university syllabi, scripts, and high school lesson projects focused on identifying and combating disinformation. The platform's goal is to strengthen social resistance to false information, especially among younger generations of Polish students and pupils. The PCD is a proposition of a nationwide action model that could ultimately cover all levels of school and academic education.

2 Counteracting Disinformation at the Individual Level

It is difficult to imagine completely eliminating disinformation from interpersonal communication, regardless of its carriers. Disinformation is, in a way, a natural phenomenon as many people spread it willingly because they want to stand out, hide something, gain something, be the first, create, or belong to a group. They often live with prejudices, think in a stereotypical way, and encounter politics or businesses that are abundant in manipulation (*The complex*, 2021). Besides these general factors, there are others that contribute to or limit the spread of disinformation, such as:

- Demographics. Age plays a key factor, as older people are often more credulous and less likely to verify information.
- Source. This involves considering the qualifications of a source. For instance, celebrities may post information while lacking competence in the area of knowledge.
- Level of education. Although more education does not guarantee the avoidance of disinformation, it promotes awareness of threats.
- Increased mental capabilities. Intelligent people are more prone to knowingly lie as it often requires more complicated mental structures than telling the truth.
- Personal traits. These traits can include sympathy for people, caution, introversion, and independence. Each of these traits individually informs people's susceptibility to disinformation or their propensity to disseminate it.
- Tendencies to analyze. People who are equipped and ready to analyze usually favor prudence in action in the information space.
- Emotionality. Those who are more emotional are at greater risk for impulsive actions.
- Morality. A person's morality is an important source of motivation for behavior in the information space.

Possessing an elementary psychological knowledge and a certain range of social competences, an individual can relatively quickly develop the necessary habits of verifying facts and not succumbing to easy instincts. Media education is needed to fill the gap in the ability to navigate safely in today's information environment at every stage of life.

3 Informational Health and Safety

Applications are also created for media education and countering disinformation in public-private partnerships, including international ones. *Breaking Harmony Square* was developed as part of an American-British-Dutch collaboration to teach recognition of manipulation techniques (Carr, 2021). In Poland, educational tools, including games for younger internet users, have been developed by entities such as the Demagog Association (*Platforma*, 2022). These training tools and games help individuals learn how to detect trolls (Aneja & Ifraimova, 2021) and how to distinguish a propagandist and a disinformer from a journalist. They also raise awareness of other risks and threats, such as the traces left on the web by clicking (Stradner & Agrawal, 2021), how trackers build a user profile based on the history of visited websites, how algorithms create information bubbles, and why content on Facebook is individually curated (Ressa, 2016).

In the interests of cybersecurity, computer and smartphone users should follow rules such as protecting passwords and changing them regularly, having up-to-date anti-virus software, and adhering to software license rules. In the information environment, to which we gain access through these tools, we must also adhere to certain rules and develop certain habits discussed in greater detail in the previous section. These rules and habits can be designated the hygiene of behavior in the information space, calling for users to:

- Check the credibility of the sources. This includes the platform, internet address, and contact data. The data and address may be intentionally crafted to look like recognized media.
- Check the credibility of the authors. This includes their previous publications, posts, and profile.
- Take time to be acquainted with the whole and verifying the integrity of the text. Take note of whether it contradicts itself, whether it is in other sources, and whether it is a satire.
- Verify the publication's date and chronology. This includes paying attention to photos or images for distortions.
- Cross-reference information in other sources and with experts.

- Notice flashy titles. This involves checking not only the source but also the quoted interlocutors and experts.
- Verify linguistic correctness.

In other words, information should be approached with skepticism and critical thinking, verifying sources, distinguishing facts from opinions, and being open to arguments with people of different views. It is also worth investing in access to reliable media and the content they provide and referring to expert studies in media education.

Over time, everyone can develop the ability to recognize not only trolls and bots but also Coordinated Inauthentic Behavior (CIB). A natural next step in identifying disinformation may involve learning basic techniques of propaganda and disinformation. For those who would like to become more actively involved in identifying and combating disinformation, social media service platforms provide free tools to track it, such as those used by the Disinformation Forensic Lab (DFRLab) at the Atlantic Council in Washington. These include (Knight, 2021):

3.1 *WHOIS Searches*

WHOIS searches allow users to see who registered a domain name, when it was registered, and sometimes find registrant contact information. This can be useful when investigating suspicious websites disseminating disinformation or propaganda (WHOIS, 2022).

3.2 *Transparency and Facebook Sections*

Limited third-party tools are available to research Facebook, but the platform itself provides information about pages and groups through transparency and sections (*How to Manage*, 2022). The administrator's location data is displayed in a transparent manner, as are the rename records which allow you to check if the page has been changed or if the administrators are in the country about which they are reporting. The *About* section can contain map coordinates, cell phone numbers, email addresses, and even website links, all of which can track site operators.

3.3 *Google Analytics and AdSense*

If WHOIS searches are unsuccessful, there are other options. A quick look at the site's source code, accessed by right clicking the page and then clicking "view source code", may reveal key information. Google Analytics information is found by searching for "ua-" in the source code. This will return a number in the format ua-12345678, which can then be used for an analytics reverse lookup using tools such as DNSlytics to verify that the same analytics ID is embedded in the source code of multiple sites. A WHOIS search may then reveal more information about the site operators. You can find Google AdSense identifiers in a similar way by searching in the source code "pub-", not "a-".

3.4 *Advanced Google/Google Dorking Search Engine*

By using advanced search operators, the DFRLab can find and identify links between the author of a propaganda report and the government from which they are writing (*Exposing The Invisible,* 2022). Google Dorking, or advanced search on Google, allows the user to diversify search engines to find information that would not be easy to find on the website, such as government expense spreadsheets.

3.5 *Advanced Twitter Search*

Advanced Twitter search allows the user to check the connections between two or more accounts on that platform, if they interacted, and if so, what topics they covered. Advanced Twitter Search also allows for bypassing the limit that prevents users from viewing over 3,200 tweets posted on an account's timeline. By using this search, the user can go back to the creation of the account by searching all tweets for a specific hashtag, keyword, or phrase, or by searching for a specific period. This allows verifiers to see if an account's performance has changed over time.

3.6 *Twitter Investigative Tools*

Twitter search tools, such as Twitonomy (*Twitonomy*, 2022), TweetBeaver, and TruthNest give greater insight into user activity. Trithionate and Twitonomy show what tools are being used to publish on the platform, which may show bot activity. Tweet-Beaver allows investigators to download the last 3,200 tweets of a user for analytical purposes. It can also search for the data of 90,000 accounts within a time span of 15 minutes.

Technology companies, media, and fact-checking organizations offer a wider range of tools and databases on various forms of disinformation. Access to some of them is paid or restricted to members of fact-checking organizations. One of the free and basic elements of the investigative toolbox is InVID. Developed as part of the European project of the same name, it aims to help journalists verify the content of photos and videos on social networks. It allows for faster and more effective verification by gaining contextual information regarding videos on Facebook and YouTube, dividing videos from various platforms into frames, reading video and image metadata, checking copyrights, and using various types of forensic filters (InVid, 2022). Similar tools are also useful for those who would like to get involved in crowdsourced fact-checking, an activity that allows organizations and the media to save costs and time. Tests have shown that there is a great deal of common sense among ordinary web users and that they have the potential to fight fraud (Dizikes, 2021).

The system of countering disinformation involves state entities, governments and their institutions, community groups, research, and the media. However, its main link is the individuals responsible for the functioning of these communities through their work. Even if some perform this work completely independently, their achievements are part of this environment's general potential. Their knowledge, commitment, individual expertise, and messaging, sometimes reaching millions of recipients, serve the purpose of unmasking, deterring, or hindering the tasks of disinformers.

Alexandre Alaphilippe, for instance, is the founder and executive director of the Brussels-based non-governmental organization EU DisinfoLab and a prominent independent researcher. His organization has a global mission to expose disinformation campaigns, raise awareness, and build civil society resilience to disinformation. Ben Nimmo heads investigations at Graphica, a company dealing with network analysis, including the use of artificial intelligence, researching internet communities, and detecting disinformation and coordinated inauthentic behavior, to discover the path of the operation of Russian influence. Peter Pomerantsev from the London School of Economics has been conducting research on information manipulation for years. In his book *This Is Not Propaganda*, he combines the story of Soviet dissidents forced to emigrate from the USSR with the history of people exposing falsehood, introducing the problem of international disinformation to millions of readers. Anne Applebaum, winner of the Pulitzer Prize, discovers truths unwanted by autocrats in all her work. Together with Peter Pomerantsev, she prepared a report on the Swedish system for counteracting disinformation. Samuel C. Woolley from the University of Texas at Austin analyzes the role of modern technologies in contemporary global political communication. In his famous book *The Reality Game*, he shows how technologies, from deepfake to virtual reality, are used to manipulate public opinion and how they can be used in the future.

Counteracting disinformation is handled by specific sections within international organizations. Individual teams and their leaders, such as Baiba Braže and Oana Lungescu, are crucial to shaping the image of NATO and building public confidence in it. Martyna Bildziukiewicz heads the EastStratcom disinformation task force in the European Union, which contributed to the launch of the Polish version of the EU vs. Disinfo databases. The team analyzes and provides tips to counteract disinformation mainly from Russia.

The undisputed global leader among these individuals is Maria Ressa, a Filipino-American journalist, writer, co-founder, and CEO of Rappler. She is also the first Filipino Nobel Peace Prize winner. Ressa has equated internet violence with real-world violence and referred to social media as a deadly game of power and money, which Shoshana Zuboff has called "surveillance capitalism". She argues that our personal experiences are being sucked into a database organized by artificial intelligence and sold to the highest bidder, with highly profitable micro-targeting operations designed to structurally undermine human will. She describes it as a behavior modification system, in which we are all Pavlov's dogs; as she notes, the experiment continues in real-time with catastrophic consequences (*The Nobel*, 2021).

Last but not least, despite repressions, Dmitry Muratov is a Russian journalist and editor of the irregular *Novaya Gazeta*. In 2021, he was awarded the Nobel Peace Prize for his efforts to protect freedom of speech together with Maria Ressa. He ended his speech in Oslo with the words: "Let us stand up and celebrate with a minute of silence for my fellow reporters and Maria Ressa, who gave their lives for this profession, and give support to those who suffer persecution. I want journalists to die old" (*The Nobel*, 2021).

CHAPTER 11

Countering Disinformation: Corporations and Civil Society

Instances of significant manipulation through social media during election campaigns have been recorded since the mid-2000s. For example, Putin's staff conducted dress rehearsals before the 2007 and 2008 presidential elections. Additionally, in his book *This is Not Propaganda,* Peter Pomerantsev describes how Rodrigo Duterte used social media to win the presidential election in the Philippines in 2012 (Pomerantsev, 2019, pp. 1–16). Following the experience of social media manipulation in the 2016 U.S. presidential election, the world's democracies began gradually taking more active measures against foreign manipulators.

1 Social Platforms: Evolution and Disinformation

During the 2017 French presidential election, a joint French and British media cluster created a platform for verification, fact-checking, and debunking of counterfeits, aimed at ensuring fair conduct that reflects the will of voters (*CrossCheck France*, 2021). Similarly, prior to the federal parliamentary elections in Germany in September 2021, a national front was established to counteract disinformation. It was supported not only by the media but also research centers, including international centers, federal government institutions, and the political parties participating in the elections.

Combating disinformation is not limited to politics and elections. In the era of digital disinformation, the biggest challenge in countering it is not just individual distortions or manipulations, or even coordinated operations of influence, but the hard-to-contain nature of social media itself. Disinformation is perpetuated by the constant presence of multiple narratives, which create and maintain information chaos.

Platforms' features overlap with the power of their corporations, and the potential benefits of dividing information giants into smaller companies remain a lofty, unrealistic goal. These platforms not only develop their empires but also produce devices, monopolizing both the channels of communication and the increasing production of necessary tools.

DOI:10.1163/9789004715769_015

To illustrate, Apple has the power to remove from its Apple Store any Facebook/Meta or Google applications that it deems harmful due to their potential for disseminating disinformation. Google will not do the same with applications created for its own smartphones, Pixel. Technological giants wield influence and economic opportunities on par with states. For instance, Meta's annual income is nearly $100 billion USD, equivalent to the domestic product of a large wealthy country. Their income will only continue to increase, not only from internet services, but also from investments in banking or cryptocurrency production. For Mark Zuckerberg, the company's importance takes priority over democracy, and the amount of knowledge Meta has about its users exceeds what they voluntarily provide. According to experts, an average of 70 Facebook "likes" provides more data than an individual's friends have, while 300 "likes" provides more data than spouses possess (Haden, 2021).

The accumulation and aggregation of user data by corporations, coupled with their increasing economic power, has resulted in a host of issues, such as the challenge of accessing service provider databases. Experts struggle to access these "black boxes", which makes it difficult to pinpoint potential risks associated with the use of these services and how they may mislead users. These databases gather information about users and their online behavior while also revealing the actions taken by platform managers to combat manipulation in web administration.

Effective measures to combat disinformation on social media platforms leave much to be desired, and without conscientious regulatory obligations, particularly from the largest states and international organizations, efforts will continue to fall short. The ambivalent approach of corporations and platforms such as Facebook, Google, or Twitter toward this issue is not surprising given that their business models prioritize instinctive user action and emotional involvement. While technological giants have taken steps to reduce falsehood and hate speech online, as well as potential manipulation and disinformation, decisive statements are often not followed up with sufficient action. Instead, they tend to be reactive and largely prompted by media pressure, research organizations, governments, and international institutions. Furthermore, smaller but increasingly powerful entities such as TikTok and Telegram demonstrate passivity in addressing the issue. The following examples illustrate the approach used by technology companies toward reducing falsehood and aggression online.

1.1 *Facebook*

- Facebook conducts its own research and activities to limit disinformation on the web.

- A monitoring board for web content was established. They publish periodic reports on moderation, warnings, content deletion, and account blocking, which includes part of the Code of Conduct for Countering Disinformation.
- Facebook collaborates in fact-checking and training for journalists and covers the cost of the trainings; the platform therapeutically redirects users from risk groups to sources advising on how to counteract extremism, which is also known as the "redirect initiative".
- It conducts tests to assess the limitation of visibility of political content in the United States, Canada, Brazil, and Indonesia. Tests are also conducted for functions like "read before posting".
- Moderators remove or suspend accounts, including those of politicians who repeatedly spread disinformation or extremist views.
- The Taliban is banned from being active on the platform.

1.2 *YouTube/Google*

- Both YouTube and Google conduct research on disinformation and methods of counteracting it (Jigsaw).
- They block the share of ad revenue from a pool of the platform (i.e., they block ads targeted at people under 18).
- They remove content published by politicians and governments that violate the platform's regulations such as misinformation on health matters or content to incite hatred.

1.3 *Twitter*

- The "Birdwatch" program in the U.S., South Korea, and Australia was launched, which allows volunteers to tag tweets that contain false information.
- Twitter started the practice of pre-bunking which is pre-empting disinformation regarding internet voting.
- Accounts of politicians misinforming about COVID-19 or spreading conspiracy theories are banned.

Reddit has blocked subgroups that disseminate COVID-19 misinformation, while other platforms have taken measures such as conducting research to counteract disinformation by creating fact-checking tools, working together with media and journalists, blocking and removing accounts (including those belonging to politicians), or applying interstitial warnings.

The conclusions resulting from the information presented above show that: (1) in the face of public pressure, proactive undertakings by platforms are dominated by protecting children and young people against disinformation and aggression as well as public health-related topics and combating extremism;

(2) there has been a noticeably more active reaction to political manipulation and extremism, particularly from the side of the largest corporations; (3) these corporations have begun to limit the possibilities of earning money from disinformation.

Companies conduct their own research on disinformation and introduce and provide free research tools in this area. They also experiment and expand their cooperation with journalists. Although these tools are insufficient to fully contain the current level of disinformation, fact-checking would be much more difficult without them.

Corporations responded swiftly to Russia's invasion of Ukraine, although their implementation of certain decisions lacked consistency. YouTube and Facebook restricted access to Russian state media in the European Union, prompting Russia to retaliate by blocking Facebook and restricting Twitter's operations within the country. Russia also cut off the broadcasting of several independent Russian-language Western media outlets, including the *BBC* and *Deutsche Welle*. Microsoft has blocked Russians from purchasing its products and services, but existing customers in Russia are still able to use them. TikTok has disabled the publishing of new posts in Russia and blocked *RT* and *Sputnik*, but other regime media outlets were left untouched. Twitter faced criticism for its half-measures, as it only limited the visibility of tweets from the regime's media in Russia while remaining open to pro-Kremlin disinformation channels.

Moderation remains the primary method for removing undesirable content on a global scale. It involves managing and administering the behavior of social media users in compliance with regulations, which include rules and guidelines that limit or block the publication of unacceptable content. Moderation may occur before publication as preventive moderation or after publication, conducted algorithmically, automatically, or by a human moderator. It may also involve temporary account suspensions or permanent account deletions from a platform's community.

Online platforms typically use mixed moderation models. In its moderation policy, Google blocks ads on YouTube targeted at users under 18 based on their age, gender, or interests. The company plans to expand the scope of blocked ads by applying additional protective filters for this category of users. Additionally, Google has introduced measures to protect the security of personal data at the request of users or their guardians, such as blocking the public ability to find images of minors through its search engine. These measures aim to extend privacy and prevent advertisements that directly prompt young people to spend money. Google is adapting to the requirements imposed on platforms by lawmakers in EU and other countries. Facebook has announced similar actions.

In one particular case, YouTube removed 15 videos of Brazilian President Jair Bolsonaro due to their COVID-19 content. Bolsonaro reacted by submitting a draft law to impede similar actions in the future, but the court blocked its entry into force. Facebook also removed accounts associated with the Russian company Fazze, which conducted anti-vaccine disinformation by attempting to bribe influential Western bloggers to criticize Western vaccines and promote Russian ones. Pakistani and Bangladeshi accounts were also used in the Fazze operation. Linguistic errors in the repeated messages exposed this peculiar laundering of disinformation.

As a result of the COVID-19 pandemic, there has been a shift in the attitudes and commitments of social media platforms in combating disinformation, leading to increased collaboration with civil society groups. Facebook has funded fact-checking training for journalists, investing around $85 million between 2016 and 2021. A powerful way to discourage counterfeiters is to demonetize their content by refusing to share in the profits of advertisements placed alongside extremist, deceptive, or manipulated publications. Google has demonetized websites that spread anti-vaccine theories through their AdSense program. Facebook has introduced the Redirect Initiative, which redirects users to websites and sources that counteract extremism (Clark, 2021). The platform is also experimenting with limiting political content in the U.S., Canada, Brazil, and Indonesia, and plans to expand this tool to Spain, Ireland, and Sweden. Additionally, Facebook is testing a "read-before-share" feature, similar to the one used by Twitter for some time.

Twitter has partnered with Reuters and the Associated Press to gain contextual journalistic knowledge about public and international affairs, allowing platform moderators to work more effectively and consciously. It has also launched Birdwatch, enabling volunteers to tag tweets they believe are misleading (Timmins, 2021). The program is currently being tested in the United States, South Korea, and Australia. Additionally, Twitter has implemented pre-bunking, a means of preventing disinformation by anticipating misinformation related to internet voting (Boman, 2020). Twitter actively monitors and responds to content that violates its regulations, detecting around 65% of such content (Dang & Culliford, 2021). It removes or suspends accounts of politicians who engage in activities that threaten public safety or health, such as former U.S. President Donald Trump and other members of Congress. Twitter employs around 1,500 moderators to manage content for 200 million daily users.

Due to the growing market for on-demand disinformation, platforms are removing more accounts that commercialize this practice. Perpetuating pro-Kremlin propaganda can earn several thousand U.S. dollars per month.

Meta has removed 6,000 pro-Russian accounts, parties, and groups from Facebook and Instagram since 2017. Twitter has removed 5,000 accounts since 2018, and YouTube has removed 2 million videos since 2019. After the invasion of Ukraine, these numbers soared. Many of the accounts removed were linked to the St. Petersburg-based troll farm, the Internet Research Agency. Russia has demonstrated the importance of YouTube in their disinformation activities through their nervous reaction to Google's decision to block the German-language *RT* channel. YouTube is therefore not defenseless against the pricing measures taken by the Kremlin in Russia (Dubov, 2021).

Russians are often behind the accounts operated by organizers of foreign disinformation on social media platforms, but there are also increasing numbers of accounts operated by Chinese, Arab, Iranian, and non-state actors associated with terrorist organizations. Despite efforts to counteract internal political or foreign manipulation, political advertising through web content remains one of the key problems. Decisions on how best to monitor and regulate this issue are still being made. Experts believe that efforts should focus on:

- Defining the ads.
- Maximizing transparency and verification by, for instance, determining who is behind the ad.
- Limiting the proliferation of advertisers of unknown provenance, especially those who appear just before the elections.
- Limiting the number of advertisements to better hold the authors accountable.
- Distinguishing ads from other content.
- Introducing election silence on online advertising.
- Applying the silence rule consistently, not only during election campaigns.
- Enforcing the rules and punishing violators (*Ten simple*, 2021).

Regulations in this area would enable better tracking of advertisements by or for foreign clients with disinformation motives. Drafts of legal acts that have emerged in the European Union and the United States suggest that increased transparency standards can be expected in this regard.

Greater proactivity is expected from corporations on issues such as engaging with researchers and providing them with greater insight into databases, algorithmic systems, and content moderation. There is also a need for more consistent enforcement of compliance with their own regulations. A problem is the lack of uniform standards and the reactive nature of actions taken only under pressure from governments or public opinion. For example, while Google limits the possibility of advertising to minors, Facebook does not (Culliford, 2021a). In fact, Facebook is even developing applications aimed at four-year-olds to attract new generations. Facebook prohibits the Taliban from being active on its platform, but Twitter does not. Reddit has blocked COVID-19 disinformation but only in response to protests from other users. Facebook works together with the journalism community on fact-checking and journalist training, allocating $84 million toward it in recent years. This is only a small portion of its income, however. These online platforms also engage in activities that either harm efforts to prevent and combat disinformation or simulate such efforts due to conflicts between public interest and their business goals and practices. Harmful or simulation actions taken by these online platforms are described below.

1.4 *Facebook*

- Facebook limits human control over the effects of algorithms in terms of moderation.
- It discriminates against weaker groups of network users.
- It has created categories of users who are attracting people and increasing the popularity of the platform, and they are given more exposure than others.
- It limits moderation or does not moderate content in rarer languages.
- Restrictive conditions are imposed for cooperation with researchers.
- Genuinely favorable conditions are created for recruitment to extremist groups, creating real health problems and rejecting proposals for adequate changes.

1.5 *YouTube*

- YouTube has seen a marked increase in harmful content recommendations.
- YouTube does not take measures to redirect to websites correcting disinformation; what's worse, it allows redirection to further disinforming sources.
- Its contextual warnings are not effective.

TikTok has been found to spread disinformation about the coronavirus and only remove extremist content after users flag it. Amazon and Spotify permit shows that spread anti-vaccine propaganda. Other platforms allow advertisers to target recipients of their ads; trade in gadgets that promote extremist and misinformation organizations; enable bypassing of bans, promote intolerance, phobias, and aggression; or are ineffective in preventing banned users from returning to the platform.

The issue of inconsistency in filtering published content has come to the forefront of public debate. An example of an arbitrary approach to moderating user-posted content is the creation of a specific category on Facebook's special XCheck user list. These users are given more freedom in their content due to their media coverage and can therefore attract greater engagement from others on the web. For instance, Facebook did not take action on a post by

Brazilian footballer Neymar featuring a naked woman accusing him of rape. This lack of action symbolizes a tolerance toward many celebrities who are also influencers.

Computer algorithms designed to prevent the automatic recommendation of users to join extremist groups on Facebook have limited success. Shocking data shows that 64% of recruitment to extremist groups is done through Facebook features such as "Groups You Should Join" and "Discover". However, Facebook management has rejected proposed models of change that could reduce this type of engagement on the platform (Hao, 2021a).

As part of the redirect initiative, Facebook is testing the Hints program that helps users find information and resources when they are exposed to extremist content. The program redirects users to information about programs such as Life After Hate. A global program like this may have more success than simply removing extremist messages. Given its position in the market and financial capabilities, Facebook should allocate more resources to fight disinformation. Expectations regarding this matter should be proportional to the company's profits or the social harm caused by unconscious or willful neglect.

Facebook also faces difficulties in moderating content in less common languages, as is the case in Ethiopia, a country with a population of 100 million people and six official languages. The company primarily relies on automated moderation in this market, which is unreliable and does little to limit escalating tensions, inter-ethnic conflicts, and even violence such as rape and homicide. Furthermore, Facebook faces significant moderation issues throughout the Arab world (Scott, 2021b), as well as in non-English speaking environments in general. This includes the Italian platform, where protection against disinformation is inadequate (Steffenhagen, 2021).

Although Facebook takes measures against large disinformation sources, it only does so in a more definitive way under pressure from governments or public opinion (Darcy, 2021). External pressure from governments and civil society is clearly insufficient as Facebook users remain disproportionately opposed to vaccination compared to viewers of the conservative Fox News media outlet in the United States (Kimball, 2021). Another issue is that online media creates content specifically for its potential popularity on Facebook, tailoring it to the platform's mechanisms of popularity (Hagey & Horwitz, 2021) at the expense of reliability.

With billions of users and exponentially more publications, YouTube's removal of a million fake COVID-19 posts (Solsman, 2021) is merely symbolic. On TikTok, 80% of extremist content is removed only after user intervention. In reality, after clicking on a COVID-19 disinformation page, users are not redirected to pages that correct disinformation but to other disinformation sources

(Sweet, 2021). Even after detecting and identifying misinformation or influence operations and their perpetrators on Twitter, some of them still persist.

Platforms could enhance their efforts to prevent misinformation and disinformation not only by removing problematic content but also by providing contextual alerts to their users. Such alerts could inform users that the information they are about to read or watch may be unreliable. Instead of redirecting users to the page containing falsehoods, the alert could break the link and prompt the user to make an informed decision about whether to proceed. Contextual warnings are frequently ignored by users, but warnings that interrupt reading or viewing tend to be more effective.

Researchers at Harvard Kennedy School have noted that platforms often have a narrow understanding of the fight against disinformation, limiting their approach to fact-checking. At the same time, they ignore the broader political and cultural context of the messages. Representatives of weaker and minority groups are often the targets of xenophobic and disinformation attacks due to willful ignorance of the issues. Counteracting disinformation should therefore not only concern facts and recognizing falsehoods or hate, but also examine power structures that favor disinformation, the functioning of technology companies, state agencies, and the entire media and information environment. The combined force of these factors saturates the web with anti-racist or anti-Islamic content in the context of the fight against terrorism and promotes the superiority of the white race and nationalist attitudes. The authors postulate a multidisciplinary approach to research on falsehood and disinformation that would allow for a wider inclusion of knowledge about history, political economy, and other social sciences in contemporary research on information space and media platforms (Kuo & Marwick, 2021).

Facebook's Oversight Board has been criticized for not enforcing its own recommendations on content moderation and automation of online operations. While it has made many unprecedented declarations that are consistent with those postulated by civic groups, it does not implement them when doing so exposes the company to excessive costs. The challenge, therefore, is to find a balance between freedom of expression, the extent of interference, and ultimately, the business model of social media platforms. Critics suggest a more scrupulous implementation of the council's recommendations through increasing platform transparency and better assessing the impact of algorithms on humans. They propose setting up multi-environmental social media boards to moderate and interfere with content (Kayyali & York, 2021). This approach is partly followed by designers of new solutions within the European Union, which will be discussed further down.

The true change that is expected in social media's fight against disinformation will not be achieved solely by moderating and deleting posts or blocking accounts but rather by a fundamental shift in the nature of the platforms' operations and their business models. They must prioritize social good and their own responsibilities over profits, either by their own choice or through government regulation. While moderation discussions are undoubtedly necessary, they are becoming increasingly more of a distraction from the heart of the problem, which the platforms find convenient. By focusing on moderation, these platforms are dealing with symptoms rather than addressing the disease itself (Melford & Rogers, 2021).

Improved cooperation between the scientific community, researchers, and social media platforms could help navigate the dilemmas and paradoxes described above. However, Facebook's restrictive terms of cooperation and blocking of researchers from accessing data, including the termination of their accounts, hinder progress (*Facebook*, 2021; Kaye, 2021). The company's aggressive stance against attempts to influence its operations is illustrated by Louis Barclay of *Slate*, who noted the threat of legal action the platform levied against a scientist who created an application allowing users to delete content from their timeline (Barclay, 2021). Unequal enforcement of regulations is also a problem, with content in English being removed but kept in other languages such as French. Inconsistencies in enforcing regulations and applying internal report recommendations complicate matters further.

The new negative trends and phenomena provoked by the coronavirus pandemic are well illustrated by the case of Telegram, one of the most popular platforms among Russians. Pavel Durov created it, just like the most popular social networking site among Russians, *Vkontakte,* but he was forced to sell Telegram to Kremlin-obedient oligarchs. The platform, once banned in Russia, was also used by Russian state bodies, politicians, and officials. During the presidential elections in Moldova in 2021, it was the main platform for disinformation targeting democratic candidates.

The popularity of Telegram has grown significantly in recent years, and it has become a platform for various activities, including extremist content and disinformation campaigns. According to Pipe (2021), Telegram has become a haven for extremists due to its lack of moderation policies. Additionally, its popularity is increasing in countries such as Germany and Spanish-speaking regions (Loucaides, 2021). As of July 2021, Telegram had 550 million users, making it a significant source of mass disinformation (Talant, 2021). Due to its accessibility and popularity in Russia, however, it can also be an essential platform for sharing accurate information about issues such as the war in Ukraine, used by Ukrainian expatriates in particular.

One unintended consequence of restrictions on open platforms is that users migrate to encrypted platforms. This leads to an increase in the popularity, income, and influence of the encrypted platforms in the market and among users.

Although corporations cannot be held responsible for all wrongdoing, it is clear that despite efforts to combat misinformation, spreading false information on social media is still profitable and even encouraged. With an annual revenue of $400 billion, Amazon could easily refuse to sell books that promote anti-vaccine ideology, while Spotify has no need to make money off podcasts that question the efficacy of vaccines. Other large companies should also avoid financing propaganda, such as ads on Belarusian state television. Additionally, corporations' claims that they do not sell personal user data to advertisers is hypocritical as it is widely known that data transfers allow advertisers to precisely identify and target recipients (*What Does*, 2021).

In addition to their failures in combating disinformation, these platforms also hold a monopoly on information. For example, in the Philippines, 96% of residents have Facebook accounts and rely on the platform for daily news and information. Meanwhile, Facebook's CEO, Mark Zuckerberg, still maintains that social media is not mass media, thereby avoiding regulations that traditional media outlets are subject to. Legal regulations enforced by states and international organizations are necessary to hold these platforms accountable. Former employees and whistleblowers, such as Frances Haugen, have revealed that the company's leadership has rejected proposals for changes aimed at protecting users' well-being, such as reducing their time spent on the platform. However, Haugen remains a loyal representative of the industry and opposes the idea of splitting up Facebook or amending U.S. media laws to hold social media platforms accountable for content. Instead, she advocates for changes in algorithm design, such as prioritizing chronological timelines and reducing emotionally charged content, which could lead to positive changes in content recommendation and user experience on these platforms (Hao, 2021a).

Some problems posed by platforms are easily identifiable and simple to deal with, such as tackling the posting of commercial ads with anti-Semitic, xenophobic, or homophobic content that promote intolerance and aggression (*Ad-funded*, 2021). However, other corporate misinformation offenses will require more complex countermeasures, more time, and more effort from both the platforms themselves and regulators.

The actions of corporations and their representatives, including former employees, should be taken seriously but also approached with caution. Their lobbying networks in Western countries and the European Union are powerful and reminiscent of how the tobacco and energy industries have protected their businesses while posing as champions of new solutions.

Global Disinformation Index experts have proposed various solutions for social media, including general and specific, voluntary or compulsory, political or legal measures. Some of these proposals have already been introduced by the European Union and its member states, but they primarily relate to the situation in the United States:

- Make platforms more accountable for spreading lies.
- Identify notorious disinformers and ban them.
- Use content moderation to raise awareness of falsehood and truth among users.
- Use awareness of falsehood and truth to collectively flag false information so that algorithms can identify lies more effectively and efficiently.
- Increase researchers' access to data.
- Offer platforms forms of limiting liability for the appearance of illegal content on them in exchange for researchers' access to aggregated data about users, their behavior, and methods of counteracting prohibited acts by platforms.
- Set up independent expert councils to review research spending on platforms.
- Demand anti-monopoly legal solutions.
- Create a global institution for cooperation and discussion on internet governance.
- Appoint statutory bodies to supervise the fulfilment of obligations.
- Offer platforms protection against liability when they fully comply with content moderation provisions.
- Split the largest companies, starting with Facebook.
- Regulate platform interoperability issues.
- Follow the example of the EU's proposal to impose stricter penalties.
- Create the possibility of running joint programs with the largest platforms to increase innovation (Decker & Boucher, 2021).

Regardless of the voluntary limitations on a global level or the ones imposed by regulators, the issue of hatred and disinformation will put the corporate business model to a significant test by new legal acts that are being prepared in the European Union, which are discussed in the next chapter. Additionally, actions have been initiated in the UN system regarding online regulation that is modeled on the UN's global process of preventing climate change.

2 Fighting Disinformation: Civil Society

The system of power control must be based on truth, just as power itself must be based on truth and recognition of the values and indisputable things on which there should be a consensus. Disinformation campaigns strive to create a world of relativized values, chaotic reality, and a sense of uncertainty in people. This is why they target major democratic institutions such as elections, human rights, social and international solidarity, freedom of speech, and truth itself. Unlike with taxation and low-cost registration, there are no havens in this regard. For example, Malta was shaken by the murder of investigative journalist Daphne Caruana Galizia, who was tracking down corruption in the government and its connections to the business world. Everyone, including the Prime Minister of Malta, opposition representatives, and journalists involved in the investigation, reported hacking attacks and falsified emails, leading to information chaos and doubts about whether the truth behind her murder would ever be revealed (Malta: Journalists, 2021).

While some may argue that the potential for online violence is not high and that its impact is primarily individual (Van Dongen, 2021), the fact remains that the illegal annexation of Crimea and Russia's incursion into eastern Ukraine included a mass-scale disinformation component. By February 2022, the Russian-Ukrainian war had resulted in 14,000 deaths and tens of thousands of injuries. Once a full invasion began, at least twice as many were killed and injured in the first month alone. Therefore, while tragic events like the Capitol riots may be the result of independent, irresponsible actors, the impact of disinformation can have far-reaching and devastating consequences on a larger scale.

Disinformation is often propagated by media influencers who exploit human phobias for financial gain. Proposed measures must consider whether exposing, stigmatizing, and criticizing these individuals through journalistic investigations, social pressure, and the risk of losing credibility and profits may be effective deterrents. Similar methods used to combat hate speech, conspiracy theories, and online extremism should also be investigated to prevent foreign interference. However, several questions and dilemmas arise, including: How can we counteract disinformation without inadvertently strengthening it? How can individual researchers and journalists tackle this issue in a world where truth has become relative and trust in journalism is declining?

Civil society plays a critical role in combating disinformation, particularly within the research, education, and media communities. These groups not only assist in recognizing disinformation but also in understanding the nature of the problem. They offer expertise, counseling, and training to public service employees, but most importantly, they educate users and stakeholders in the information space. The scope of expert initiatives engaged in this work in Western countries is so vast that preparing a comprehensive map of these organizations and outlets could be the subject of a separate study.

In the international arena, American and British universities and research centers are highly influential due to their potential, resonance, scope, and impact. Many of these institutions collaborate with experts from other countries to combat disinformation. The American Atlantic Council, for example, has specialized teams that deal with disinformation, and their analytical work is used not only by the U.S. government but also by other countries and international organizations. The Center for European Policy Analysis, RAND Corporation, and the Brookings Institution also regularly provide analyses and recommendations for governments. The German Marshall Fund of the United States (GMF) has launched the *Alliance for Securing Democracy* project, which raises awareness of the dangers of disinformation, publicizes the results of scientific research, and regularly provides summaries of narratives and disinformation activities by Russia, China, and Iran on their website, *Hamilton 2.0 Dashboard*. The GMF also created a special project to analyze foreign narratives used during the election campaign in Germany in 2021.

Europe is not lagging when it comes to disinformation measures. The network of national and international organizations and institutions dedicated to fighting disinformation is constantly growing, and the cooperation between them is becoming increasingly fruitful. In Brussels, there are many specialized think tanks and initiatives focused on analysis, fact-checking, and education, including EDMO, EU DisinfoLab, and Lie Detectors. These organizations, together with institutions operating in EU member states, form networked communities that receive organizational and financial support from the European Union. Universities are also doing significant work individually or as part of international research clusters.

In Ukraine, *Stopfake* is one of the most effective networks for tracking and revealing disinformation. It is active in many countries and in many languages. It was established on the initiative of university staff and journalism students at the Mogilev Academy in Kiev (Stopfake, 2022). Many other activities were undertaken, first by the Ukrainians themselves and their organizations, then by others, including masses of ordinary internet users worldwide. The activities were aimed at combatting Russian propaganda and information falsehoods related to the war in Ukraine. Some of these activities are unprecedented for their grassroots scale, as was the case with the group Anonymous massively attacking websites of Russian state institutions, including the Ministry of Defense and the secret services.

Selected examples of international activities undertaken in the field of countering disinformation by experts, researchers, journalists, and teachers, include:

- Bellingcat, the European Disinformation Media Observatory, the International Fact-Checking Network, all of which initiated and developed an international cooperation of researchers and journalists dealing with investigations and fact-checking.
- BBC, Lie Detectors, which focuses on education in schools.
- The Center for Countering Digital Hate, which unmasked and made a list of the 12 largest global COVID-19 related disinformers.
- The Center for International Resilience Detection, which detected a network of 350 accounts spreading powder propaganda in France.
- The German Marshall Fund of the U.S., which participated in pre-election monitoring focused on messaging by foreign actors.
- The Global Disinformation Index, which has diagnosed the most popular information portals in terms of disinformation potential in selected countries and researched the risk of disinformation and media credibility.
- GLOBSEC, which is developing a Decalogue of Transatlantic Principles for Combating Disinformation and is responsible for the regional initiative of the Alliance for Healthy Infosphere.
- Code for Africa, which is an international network dedicated to addressing technology, journalism, and fact-checking problems.
- DROG, Cambridge University, and the U.S. State and Homeland Security Departments, which facilitate international projects of training programs.
- Mandiant, which has identified perpetrators of hacking activities.
- MEMO 98, Who Targets Me, and Citizen D, which monitored election campaigns in Slovakia and Slovenia.
- The University of Oxford, University of Michigan, and Meedan, which researched and created of an algorithm facilitating the fight against disinformation on communication platforms.

The need for an integrated global approach to combating disinformation is demonstrated by the appeal that came from the milieu of European and American researchers. It aims to promote the document containing the *10 Transatlantic Principles for a Healthy Online Information Space* (*10 Transatlantic*, 2021):

1. Strive for greater transparency in the online information space.
2. Empower users to make informed decisions about their data.
3. Foster a culture of digital responsibility and accountability.
4. Minimize the spread of harmful information online.
5. Work towards timely, standardized, and proportionate rules for the digital space.
6. Support the ethical use of AI systems that embrace democracy and human rights.
7. Develop tools to increase citizens' media and digital literacy.
8. Empower civil society and the public to get involved.
9. Nurture an open space for competition to avoid monopolies.
10. Search for transatlantic solutions and beyond.

The call for international cooperation on the basis of these principles was initiated by the Slovak organization GLOBSEC. Its other regional initiative is the *Alliance for Healthy Infosphere,* which bring together think-tank centers from Central and Eastern Europe to combine their expertise and activities and counter disinformation more effectively.

The Propaganda Research Laboratory at the University of Texas at Austin, along with other experts, initiated the development of the *10 Transatlantic Principles for a Healthy Online Information Space*. Their two-year research focused on analyzing the network behavior of American propagandists working for political parties, national or foreign government agencies, or consulting or PR firms. The study found that manipulators use both coded platforms, such as WhatsApp and Telegram, as well as more open platforms like Facebook and YouTube, to influence minority voting behavior in specific states or cities. The groups targeted by manipulators include immigrant communities in swing states like Florida and North Carolina, where their voting behavior may sway presidential elections (Woolley & Sawiris, 2021).

Technological advancements in machine learning have made it possible to accelerate the tedious process of verifying information, such as fact-checking automation. Scientists from the University of Oxford have created a special algorithm that informs WhatsApp users if the message they received has been verified for authenticity (*Tackling misinformation*, 2021). This is an example of the closer cooperation between journalists, institutions, and organizations specialized in detecting disinformation, and the use of artificial intelligence. With increasingly advanced techniques and the cooperation of research centers, journalistic and expert circles are joining the fight against disinformation in various contexts, including elections, media education, tracking and exposing perpetrators, and studying the relationships between disinformation and hackers. They are also creating maps of disinformation media and considering how to redesign the functioning of the internet and social media.

The following points synthesize important areas of activity from selected initiatives in the prevention and fight against disinformation.

2.1 *Elections*

Initiatives like MEMO 98, Who Targets Me in Slovakia, and Citizen D in Slovenia not only monitor election campaigns but also initiate legislative projects. Despite political forces defending themselves against transparency, their experiences and work show the scale of challenges regarding transparency and fairness of electoral processes. In many places in the West, parties, politicians, and the government's restraint leaves room for abuse in these matters. However, these challenges can be mitigated by international pre-election monitoring by global think-tanks. These include organizations focused on messaging by foreign authors appearing in the activities of broadcasters which may show further interference in the course of the campaign. The GMF *Alliance for Securing Democracy* project, mentioned above, contributed to observing the 2021 elections to the Bundestag while detecting and reducing foreign interference and disinformation.

2.2 *Media Education*

Research carried out by the RAND Corporation on countering disinformation has shown that most analysts propose changes in government policies, particularly to improve media education. While governments and supranational organizations like the EU can provide support, without increased participation from civil society, research, and journalistic communities, media education will likely remain inadequate in many countries. These communities possess valuable expertise and specialist skills that can be shared with educators. Initiatives such as Lie Detectors and Demagog.org, often created as a result of grassroots community efforts, have demonstrated effective ways to do this, including by incorporating games and play into mainstream education.

The potential of these lighter forms of fighting disinformation has been recognized not only by Cambridge University psychologists and Dutch activists, but also by the U.S. Departments of State and Homeland Security, who collaborated on the aforementioned game, *Harmony Square*. In the game, the player assumes the role of the "disinformation director" whose task is to sow discord and disrupt social harmony. The game, developed as part of an international project, uses the previously mentioned approach of psychological inoculation. The game creators also developed similar products for children and teenagers.

2.3 *Attribution*

Bellingcat is known for conducting extensive investigations that go beyond exposing acts of disinformation. Their investigative potential, operating model, and ability to mobilize international cooperation make them a leading civil society institution. They have been instrumental in exposing the organizers of influence operations, including international disinformation campaigns like relating to the shooting down of Malaysian Airlines flight MH17 and attempts to poison Sergei and Julia Skripal. They have also uncovered manipulation of information during hostilities, such as those conducted by Russia in Ukraine. Other organizations, such as the Center for Countering Digital Hate, have compiled lists of the most influential and harmful global manipulators of information, including those spreading misinformation about the coronavirus. The Center for Disinformation Resilience (CIR) recently detected a network of 350 accounts spreading pro-China propaganda in the French-language information space (Carmichael, 2021), prompting YouTube and Facebook to take more decisive action against their infodemic activities.

2.4 *Mapping Disinformation*

The Global Disinformation Index (GDI) partnered with a Malaysian organization to identify the most popular portals in Malaysia with potential for disinformation by assessing the likelihood of encountering falsehoods (Media Market, 2021). This approach holds promise for detecting and preventing disinformation on a large scale. GDI has also conducted a media credibility study in other countries, such as Brazil, where half of the tested media were determined to have a high or very high risk of disinformation (*Disinformation Risk*, 2021).

Studies by the NATO Center of Excellence for Strategic Communication in Riga and the Political Capital Institute in Budapest provide a specific catalogue of media in Central and Eastern Europe, including Poland. These reports reveal an interesting pattern: media outlets with more "Balticness" in their names in Estonia, Lithuania, and Latvia and more "Polish", "national", or "independent" in their titles in Poland are more likely to have connections with disinformation activities, mainly by Russia (*Where to look*, 2017).

2.5 *Advanced Techniques*

States that organize disinformation campaigns often outsource their activities to third parties to conceal their involvement. However, investigative techniques and metadata analysis can reveal such connections and detect the direct actors responsible for creating, for instance, a network of thousands of bot accounts or operating with an equally large number of accounts pretending to be real people within coordinated inauthentic behavior. Perpetrators also use programming techniques that confuse platform algorithms by simulating the authenticity of the account. Coordinated inauthentic behavior can be coupled with manipulation using genuine accounts and people who duplicate manipulated content, even in mainstream media. Identifying such activities requires not only knowledge of basic computer techniques but also advanced techniques, broader knowledge, and detection tools.

In disinformation operations, particularly the most dangerous ones, the key element may be hacking into email or social media accounts and then manipulating the publications of the stolen content. Methods of identifying the perpetrator of such hacking activities, including operation "Ghost Writer", which targeted Poland, were discovered by Mandiant, the analytical arm of the global cybersecurity company FireEye. Its reports indicated location certificates, specific use of Tactics, Techniques, and Procedures (TTPs), and phishing via e-mails when senders impersonate cybersecurity experts (Roncone et al., 2021).

2.6 *Cooperation*

International cooperation between researchers and journalists involved in fact-checking, pooling forces and resources, and initiatives like Bellingcat demonstrate that countering disinformation is not only time-consuming and labor-intensive but also costly. Networking is therefore an effective way to share resources and tasks. It can be assumed that joint efforts of researchers and journalists contributed to greater transparency in the presidential elections in France in 2017, the presidential elections in the United States in 2020, and the disclosure of disinformation, leading to the victory of a democrat in Moldova in 2021. For years, Germany was criticized for its passivity in responding to Russian disinformation, and in the parliamentary elections in 2021, many organizations in Germany finally cooperated to address the issue. While criticism certainly played a role, increased awareness of threats in the world of politics and government structures, intensified by pressure from civil society, also contributed to the actions taken.

Community cooperation, though most visible in Europe and North America, is beginning to extend to other continents. Code for Africa is the largest network of international cooperation focused on solving problems of media technology, journalism, and fact-checking (Knight, 2021). This is paramount in an interconnected and global world where continental, regional, and national weaknesses are eagerly exploited by disinformation organizers.

2.7 *A New Model of the Internet*

Researchers are increasingly focused on managing online platforms and their impact on people, regulatory needs, and the role of actors in this process, largely due to the problem of disinformation. Francis Fukuyama, an American scientist, has examined the idea of rebuilding and managing the internet in the modern era. He concluded that the power of platforms is so significant that they can determine the outcome of an election, and he advocated for reducing this potential. To achieve this, he proposed establishing middleware that would allow users to have greater control over the content they consume. Platform custodians, or librarians, would be at the center of this process, ensuring that information is based on knowledge and facts. Such structures should be integrated by the platforms themselves to avoid direct state interference (Fukuyama, 2021). This is an attempt to reconcile the freedom of the internet with consequences similar to regulatory effects. While implementing Fukuyama's ideas would likely lead to fragmentation in the network, it would give most users a chance to choose reliable sources more often than in an environment where the choice is left to the platforms through algorithms.

Samuel Woolley suggests that, to protect people from existing threats and problems, a new model of democratic internet management must be created. He believes that artificial intelligence is a challenge comparable to the control and non-proliferation of weapons of mass destruction (Woolley, 2021). The author proposes specific education for scientists, researchers, and data processing specialists to consider the potential effects of their implementation during the development of new digital solutions. Meanwhile, industry leaders like former Google CEO Eric Schmidt suggest in public statements that, similar to nuclear weapons, once the genie is released, it cannot be squeezed back into the same bottle. Even if the threat has passed its tipping point, however, the passivity of potential victims would only accelerate the emergence of further problems.

Since the mid-2010s, especially after the 2016 US elections, disinformation has garnered significant attention, resources, and cooperation. However, constructive critics argue that these efforts are often short-term and focused on pre-election periods. The actors involved in these efforts often do not work together or share the results of their work. Additionally, the mediatic nature of the topic means that lower-quality products can easily infiltrate the information space. Fragmentary research on the short-term effects of influence operations and overly simplistic attributions of responsibility undermine the effectiveness of countermeasures. To address these issues, there is a need for more funding, coordination, and development of verified research methods. This applies to Poland as well, where the Polish Institute of International

Affairs (PISM) and the Center for Eastern Studies (OSW) are among the centers with international reputations conducting and publishing regular research on disinformation in international relations.

3 Disinformation and Challenges for the Future of Journalism

Journalists are at the forefront of the fight against disinformation. The future of this profession depends on the credibility of the media and, more broadly, on how to make people willing to pay for reliable information. Traditional media is often seen as part of the establishment, making it difficult to reach younger audiences. The European Broadcasting Union emphasizes two key elements in countering disinformation: content moderation, which includes not only deleting posts but also addressing user complaints and ensuring transparency in their resolution, and law enforcement. While these are important steps, they represent a narrow approach that does not consider the nature of social media and its impact on the information environment. The enormity of the work undertaken by the media and journalists cannot be overstated, and they face significant challenges and threats. Research has shown that journalists covering COVID-19 were under the same level of stress as healthcare workers during the pandemic (Osmann, 2021). Despite this, their commitment to sharing the truth during the pandemic has led to a clear increase in trust in credible media in many countries, although unfortunately not on a global scale.

At the same time, fighting against disinformation requires determination and even boldness, which means providing help to those who dare to fight it. Jessikka Aro is one of the best-known examples of a journalist who felt alone in the face of pro-Kremlin aggression on the internet and decided to leave Finland. The journalist became a target of massive trolling, including threats from pro-Kremlin circles, against which she felt the Finnish authorities had provided insufficient protection.

Hate affects most journalists who deal with disinformation, and thankfully their work is highly appreciated, as in the cases of Maria Ressa and Dmitry Muratov, who were awarded Nobel Prizes. However, besides the journalism community, governments also have a primary responsibility to protect journalists' independence and security. Free media, truth-based science, and education are the cornerstones of democracy, and protecting them is crucial. The fight against disinformation starts with recognizing what it is. People and institutions, even those with greater authority, can make mistakes.

There are numerous examples of established media and state institutions responsible for public affairs disseminating misleading information. For instance, the renowned German daily *Handelsblatt* repeated unverified information about the low effectiveness of the Oxford-Astra Zeneca vaccine among the elderly and erroneously stated that the U.S. epidemiological authorities had made premature decisions to cancel the obligation to wear protective masks.

In January 2022, when the tension around the Russian-Ukrainian conflict grew due to the expansion of Russia's military forces on the border with Ukraine and aggressive war rhetoric, Germany became the subject of mass criticism for what it did not do. They did not refuse to fly British planes with military equipment for Ukraine over their territory as they did not receive requests for permission.

Journalists often face the challenge of avoiding the spread of disinformation while maintaining their credibility. One approach is to listen to the concerns of the audience and prioritize transparency and authenticity in their reporting. Artificial intelligence can also play a role in improving journalism, but it must be used appropriately. Journalists must work to engage with audiences who may feel excluded and avoid reinforcing disinformation while reporting on it. It is widely agreed that simply negating false information without providing context or explanation can inadvertently spread it further. This is a tactic used by disinformers and some politicians who prioritize exposure over accuracy. The guiding principle for journalists therefore be "first of all, do no harm".

There is a growing sense of the need to revise certain complex concepts such as journalistic neutrality in light of the prevalence of lies and fake news. During the tenure of former U.S. President Donald Trump, one of the most prolific producers of fake news in recent years, the old truth that what the head of state says is news had to be revisited. For example, American journalistic dilemmas surfaced prominently regarding attempts to amend the electoral law to be called publicly electoral restrictions, as the Democrats wanted, or fairness laws, as the Republicans wished (Hobes, 2021). In other words, comparing the incomparable and attempting to forcefully look for arguments to balance criticism and offer the rostrum to liars is not an effective approach. Journalists need to consider how not to fall into the trap of symmetries. BBC journalist Rebecca Skippage (2021), head of the monitoring team, evoking the thoughts of Hannah Arendt, directly referred to such a war of narrative: "Talk in an engaging way, like the bad guys". This is one of the main messages for journalists who wish to fight false information.

There are three basic premises and tasks from the point of view of journalistic effectiveness in counteracting disinformation.

The first premise is responsibility and trust. The media and journalists remain at the forefront of the fight against disinformation. For many of them, there are important, often personal, issues at stake, including survival in the profession. The state should support them in these efforts, finance media education with their participation, guarantee reliable public and private media independence, and maximize access to recipients. The main challenges posed by misinformation and the temptation to engage in unethical journalistic practices are concentrated within the media industry, acting as a lens to magnify these issues. The controversial self-promotion tactics employed, for instance, by the UK's *GB News* network, while technically permissible, undermine public trust in mainstream media more broadly (Orr, 2021). Additionally, pro-Kremlin trolls exploit the fundamental right to freedom of expression in the West by leaving comments on articles published in major Western news outlets. These comments are then used as propaganda in Russia to support the Kremlin's disinformation campaigns. This is a new method of Russian-controlled disinformation that has escalated in recent years as social media platforms have intensified their efforts to combat it.

It is therefore essential to have reliable content moderation on the internet, not only for social media platforms but also for traditional media outlets. Moreover, traditional media should be cautious when downgrading their content standards on social media to increase their reach at the cost of their message's reliability.

The issue of credibility and trust in traditional media is also linked to how some social groups, particularly younger generations, perceive them as an "us versus them" scenario. The "us" represents the rebellious, alienated, and disadvantaged, while the "them" represents the wealthy elite. To build trust, journalists must engage directly with these groups by participating in media education programs in schools and local communities.

The second premise is fact-checking, which involves complex activities that require cooperation and time. Dozens of media outlets and hundreds of journalists have collaborated on some international journalistic investigations. From an efficiency and credibility standpoint, therefore, it is particularly important to join forces, pool resources, and share information. The trends in media development or the crisis in traditional media will make fact-checking one of the key trends in journalism. Its main goal is to improve the quality of public debate, which is contrary to what information manipulators want. This includes verifying the statements of politicians, officials, or other influential people who find their way into the public sphere. Fact-checking is necessary, but it is not sufficient on its own. Its corrective function is implemented *post-factum* when the damage has already occurred. Research has shown that the human mind is often resistant to late corrections, and exposing disinformation has varying degrees of effectiveness depending on the recipients.

Fact-checking can help to reduce the impact of manipulated narratives and disinformation on people's attitudes and beliefs. However, it may not always change their support for a particular political party. In such cases, pre-empting disinformation by providing civic education and encouraging people to identify unreliable or biased sources may be more effective. Renowned research centers like the NATO COE in Riga or the Global Disinformation Index can assist in analyzing and mapping these problems.

To date, fact-checking has not been profitable, which has limited the resources available for its development. However, stakeholders such as Google and Facebook do pay for outsourcing, with Facebook paying around $2 million in 2019. Fact-checking services like Full Fact in the UK or Liberation in France receive around $200,000 per year from these stakeholders.

In recent years, mainstream media has invested more resources into fact-checking. Agence France Press's (AFP) Fact Check, which started as a one-person unit in 2017, has grown into a team of 120 people working in 24 languages and 80 countries as of 2021. The practice of fact-checking is evolving into a more digital and efficient process. In fact, fact-checking played a significant role in the context of the 2021 parliamentary elections in Germany. Funded by a Google grant, DPA and FaktenCheck21 trained 600 journalists from 100 media organizations. Despite the challenges facing this field of journalism, the public demands its growth and development (Scire, 2021)

Despite what some people claim, fact-checking does not have to be counterproductive (Grabmeier, 2021). In fact, by denying untruth, fact-checking, if done properly, can help to dispel falsehoods without amplifying them. That is why disinformers in Spain have attempted to discredit journalists and fact-checking itself by impersonating fact-checkers in the eyes of confused recipients (Loucaides, 2021). Crowdsourced fact-checking, which involves engaging ordinary internet users to verify information, has also shown its strength as an innovative method. However, there is a risk that this activity, while filled with good intentions, may not necessarily involve professionals with the necessary skills.

The third premise is media education. As previously mentioned, media education is a neglected subject in schools in the vast majority of countries, including many in Europe, due to the lack of curricula and teaching competencies. To address this, the media community and journalists can be involved in training educators or delivering the programs themselves. Funding for such activities could come from a portion of tax revenues, such as an audio-visual media tax, and combined local and regional government resources. In the UK, the BBC's *School Report* and in Finland, journalists' *Faktana, kiitos!* programs involve a growing number of students. Similar initiatives in France, such as *Entre les lines* in partnership with *Le Monde* and AFP, have also gained traction. International efforts to pool resources have also been successful, such as LieDetectors, a Brussels-based non-governmental organization that conducts media education programs in schools across many EU countries, including Germany and France.

CHAPTER 12

Countering Disinformation: State and International Level

In recent years, many governments have taken more decisive steps to counter the increased spread of disinformation on the internet. In the United States, antitrust proceedings are underway, and proposed legislative changes would hold platforms liable for certain content published on their sites. In the United Kingdom, the regulator requires platforms to protect data and create safer conditions for underage users. The EU has a directive on audiovisual services, and work on a new legal act on digital services has been completed to create further regulations in the Member States and partner countries of the European Economic Area. The COVID-19 pandemic and Russia's invasion of Ukraine have highlighted the need to establish a new, more effective framework for the functioning of the information and internet environment, including social media platforms' efforts to counter disinformation.

1 Dilemmas and the Need to Fight Disinformation

Despite intentions to protect freedom and the right to reliable information, some measures and proposals for combating disinformation face resistance in democratic societies. Facebook's decision to ease its policy on hate speech, however, faced criticism in light of Russia's hostilities in Ukraine. The essence of the social contract lies in finding a balance between security and freedom, and in this paradigm, it is not the state but citizens, civil society, journalists, media, academic and research institutions, and internet platforms that should be on the front line of the fight against disinformation. The Canadian government presented such an approach in 2019 – one that focuses on equipping citizens with the knowledge and measures needed to counteract disinformation (Carvin, 2021). Web regulation does not need to contradict freedom of speech, just as there is no contradiction between driving a car and having a license, using seat belts, or obeying speed limits.

The 2019 report prepared by the OSCE, paradoxically on behalf of Russia, discusses the application of international law and standards that ensure freedom of expression. However, these standards are limited in regard to the rights of other people or the interests of state security, highlighting the tensions between freedom of expression and restrictions. The issue is not whether restrictions are allowed at all but when and to what extent they are allowed (*International Standards*, 2019). At the same time, calls for a narrow application of restrictions on freedom of expression echoed in UN and OSCE declarations often fall on barren ground in many Member States that subscribe to them, including Russia.

Freedom of expression is enshrined in Article 19 of the Universal Declaration of Human Rights, Article 19 of the International Covenant on Civil and Political Rights, and Article 10 of the European Convention on Human Rights. Article 20 of the Covenant prohibits the promotion of war and hate, including based on race. Furthermore, the European Court of Human Rights has ruled that freedom of speech includes offensive or shocking statements, but limitations must be clearly defined by law and must achieve the objectives set out in Article 19 of the Covenant (International Pact, 2021). It is also important to note that false information is considered a violation of the law in various regulatory contexts.

In many countries, penal codes prohibit the dissemination of false information if it causes damage or violates civic, public, or state interests. These prohibitions are usually narrowly defined, however, and sometimes only apply to lies that are disseminated about electoral candidates. Regulations that limit the freedom of expression vary across different countries, with constitutional protections accompanied by statutory solutions and political doctrines. For instance, in Canada, false information that infringes on the public interest is criminalized. Yet even the denial of the Holocaust was not recognized by the Supreme Court of Canada as an unlawful act; it was permitted under the right to freedom of expression.

The Greek penal code provides for up to two years in prison for the dissemination of false information, while the broadcasting law obliges the media to be objective. In Croatia, false market and capital market information are penalized under the criminal code. Kazakhstan has introduced a penalty of two to five years for the dissemination of false information that threatens state security. However, while this activity warrants general analysis, the broad nature of the catalogue of such interests requires special attention. On the one hand, it reflects the changing nature and growing number of challenges to national security, but on the other hand, it may provide a gateway to repressions that authorities interpret loosely as undesirable.

The Lithuanian constitution guarantees the right to freedom of expression, access to information, and information. Nevertheless, it also explicitly states that it does not grant approval for disinformation. This provision has been adopted in the law pertaining to public information access. The

Radio and Television Commission is authorized to penalize broadcasters for instigating hatred or war, but such sanctions require approval from a court. Slovakia has also mandated the obligation of impartiality in its media law, which is overseen by the Media Council. The council has the power to penalize violations, subject to judicial review. In the UK, the media code prohibits false information, particularly in news bulletins aired by broadcasters. Additionally, the UK government communications office sets fundamental standards for content broadcast. However, most of these regulations were established prior to the digital age and the West's current challenge of external disinformation.

To counter new threats, a report by the Brennan Center at New York University's Law School recommends, among others, that a federal commission be established in the United States to define rules for researchers' access to platforms. This would help safeguard users' personal data and prevent illegal targeting (Hendrix, 2021). Such an approach would provide independent and credible social control without requiring the government to take the lead. This does not imply public consent for government inaction, however. In response to the changing nature of disinformation threats, it is necessary to adapt strategies, legislation, and administrative practices. In the West, for example, decisions to limit the activities of the Russian state or Kremlin-supported media were made quickly following the outbreak of full-scale military operations in Ukraine. Estonia and Poland were the first to remove Russian stations from the programs offered by television service providers.

The examples discussed and the data presented below only illustrate a selection of possible systemic or incidental actions and do not reflect the full range of measures taken. Nonetheless, they enable generalizations and conclusions that will be presented further down in this chapter.

- Australia's courts ruled that network publishers are liable for content posted by recipients under their content.
- Canada, Germany, and Poland developed new national cyber defense strategies.
- Estonia revoked the media licenses for *Sputnik*.
- France has implemented legal regulations that impose obligations on platforms. These include storing deleted content and designating employees to contact administration. Additionally, new structures have been established to combat disinformation prior to the 2022 elections. France has also adopted a military doctrine of information operations and identified the Russian perpetrators of the "Macron Leaks" operation.
- German authorities openly accused Russia of conducting cyber-attacks on German politicians.

- Italy opened a national center for combating disinformation.
- Luxembourg refused to grant a television license for the German-language channel *RT*.
- The Netherlands expelled spies and the authorities publicized the attempted hacking attack on the OPCW in connection with the attempt to poison the Skripals.
- Slovakia's internet regulators called for civic organizations and citizens to send drafts of projects to fight disinformation.
- South Korea strengthened public-private partnerships and support for fact-checking.
- Sweden and the United Kingdom created a guide on countering disinformation for officials.
- Switzerland's embassy in Beijing requested the removal of articles on COVID-19 that falsely cited a Swiss scientist.
- The United Kingdom created a law to protect the personal data of minors; it also proposed the appointment of a commissioner for combating disinformation in Scotland.
- The United States has taken several actions, including establishing the State Department Center for Global Commitment for diplomatic measures and sanctions. The U.S. Cyber Command launched a preemptive attack against the Internet Research Agency in St. Petersburg before mid-term elections in 2018. An initiative was also carried out to amend the Internet Act and hold social platforms accountable for disinformation content posted on their platforms. Other legal bills were introduced, including an act on fair advertising and requiring online platforms to give users options to protect their privacy and personal data given that discontent online can lead to violent protests in the streets. Lastly, mandatory media education was introduced in Illinois.

Based on these and other examples, it is possible to catalogue the actions taken relating to countering disinformation both as a general phenomenon and in its international manifestations.

1.1 *Building Structures*

The Viginum unit was established in the Office of the Prime Minister of France with a broad mandate to fight foreign disinformation in the digital domain. The mandate led to the creation of 70 positions in strategy, operational and ethics divisions. In Sweden, the Swedish Civil Contingencies Agency (MSB), entrusted with the fight against disinformation, was transformed in 2022 into the Swedish Psychological Defense Agency, with a broader mandate. In the United States, the Department of State's Center for Global Engagement was given the competence to cooperate with foreign partners. In Italy, a national hub for the fight against disinformation was created within the European Digital Media Observer network. Its center is the LUISS University in Rome, where fact-checking experts, media experts, researchers, and journalists work side by side. Structural measures of this kind are usually elements of comprehensive solutions that create the premises for a more effective fight against foreign disinformation.

1.2 *Legal Regulations*

In Germany and France, laws were enacted to combat online hate speech and disinformation. These laws served as the foundation for the drafting of the Digital Services Act of the European Union. In the United States, proposals have been put forward for new legislation, including the Fair Advertising Act, which would require platforms to publish and store data on political advertising, including its recipients and sponsors, according to standardized criteria. A proposed bill was also submitted to Congress that would require platforms to provide users with the option to restrict the use of their personal data and block algorithms that control the content they receive (Gold, 2021). Furthermore, the Compulsory Media Education Act was passed in Illinois (Eng, 2021). In Australia, an unprecedented court ruling held publishers accountable for content posted by their audience under their material. Although the effects of states' actions on regulating and combating disinformation are still limited due to the complexity of the issue, individual states and the international community can develop their own solutions.

1.3 *Cybersecurity*

Several countries including Canada, Germany, and Poland have adopted or are in the process of adopting new or revised national cybersecurity strategies. While cybersecurity strategies are important, they cannot replace comprehensive strategies and doctrines to combat the full spectrum of disinformation. An example of offensive preventive actions are the preventative attacks by the U.S. Cyber Command launched against the servers of the Internet Research Agency in St. Petersburg before the 2018 U.S. midterm elections. Other countries, such as Denmark, Estonia, France, and the Netherlands, are also developing offensive cybersecurity capabilities to prevent external disinformation and have offered these capabilities to NATO if needed. The use of such capacities is governed by the existing regulatory framework and international normative framework, as analyzed in the Tallinn Manual series of documents by independent researchers and lawyers. However, the process of intergovernmental agreements within the UN has been unsuccessful for years due to fundamental differences in the positions of Western countries on the one hand and Russia and China on the other as regards sovereignty over the Internet.

1.4 *Educating Authorities and Officials*

Instructions for these target groups in the form of extensive textbooks were developed in Sweden and Great Britain (Band, 2021). Training programs for public figures, legislative members, and executive authorities are carried out in many countries, including Poland; however, Sweden and Great Britain seem to have the most extensive programs in this regard.

1.5 *Public Warnings*

The United States Department of Homeland Security has implemented a web monitoring system to issue warnings to state and local authorities about potential public order disruptions and violent demonstrations arising from online calls. These warnings are triggered by monitoring the mood and activities of conspiracy groups like QAnon or activities similar to those preceding the events of January 6, 2021 at the Capitol (Sands, 2021). The involvement of foreign agents in inspiring such threats through social media has already been mentioned. To prevent the actions of foreign services, some countries, such as Estonia and Finland, regularly publish intelligence reports on threats, including those related to disinformation and cyber threats.

1.6 *Administrative Decisions*

Prior to Russia's invasion of Ukraine, Estonia revoked the license of the Russian news agency Sputnik. Similar periodic license suspensions were imposed by Latvia and the United Kingdom. Luxembourg denied the German-language channel RT (Lambert, 2021) a television license. In Australia, authorities ordered Facebook to cease the spread of false information on WhatsApp regarding lockdowns and store closures in Sydney during the pandemic (Taylor, 2021). Additionally, Hamburg's Data Protection Commissioner, Johannes Caspar, made decisions limiting Facebook's actions (Schaer, 2021).

1.7 *Diplomatic Actions and Attribution*

Poland reported a cyber-attack on politicians' accounts to the EU and NATO institutions. The Netherlands diplomatically and publicly amplified the attempt and mechanism of a hacker attack on the headquarters of the Organization for the Prohibition of Chemical Weapons in The Hague against the background of the Novichok poisoning of the Skripal family in Salisbury. The Swiss Embassy in China asked the Chinese media to remove deceptive content in articles it circulated about COVID-19, which incorrectly cited a Swiss scientist (Tewari, 2021). Despite these examples, states rarely take such actions to publicly attribute disinformation or cyber-attacks to the perpetrators.

1.8 *Working Together with the Public*

In Slovakia, the Internet Council has asked societal organizations to send in civic projects to counter disinformation. Long lists of similar examples can be cited, but the ones mentioned above show the spectrum of possible actions at both the national and international level. This leads to the conclusions below.

First, from a global perspective, states are reacting insufficiently and too slowly to the problem of disinformation. Only a few countries (Australia, Finland, France, Germany, Sweden, the United Kingdom, and the United States) have introduced systemic solutions. This applies to countries that are the main targets of disinformation by global state and non-state actors, but also those not in their focus, as in the case of Sweden.

The second conclusion is that, in terms of legal regulations, the most comprehensive solutions have been adopted by France and Germany, with the UK's regulations being less extensive though still significant and precursory. The United States has adopted regulations at the project level, but their adoption remains a significant challenge due to the interests of American technology giants and the political and legal culture. Human rights in the digital age present a difficult dilemma in the fight against disinformation. Proposals by American lawmakers to impose liability for falsehood on platforms such as Facebook are questioned by renowned legal experts who refer to the constitutional protection of freedom of expression.

France and Germany are examples of countries that have enacted strict laws against hate speech, disinformation, and non-transparent political campaigns on social media. This is partly due to experiences with Russia's interference with their state institutions in 2016 and 2017. However, it is also proof of the well-established determination of the societies and authorities in both these countries to counteract disinformation.

Paradoxically, while some democratic countries struggle to find a balance between combatting disinformation and protecting freedom of expression, authoritarian regimes have been using the guise of fighting disinformation to suppress dissent and control the narrative (Dang & Culliford, 2021). Countries such as Russia, China, Pakistan, Belarus, Uganda, and Nigeria have increasingly demanded the deletion of content and blocked internet platforms. Turkey and India are also sometimes leaders in these types of efforts (Paul, 2021). For instance, Uganda blocked internet access after Facebook removed pro-government accounts, while Nigeria did the same to Twitter over the deletion of a president's tweet. In some countries, disinformation has also been used by the government apparatus and even the military, as seen in South Korea (Tworek & Lee, 2021), Canada, and Jordan (Elliott, 2021a). In the Philippines, the political online disinformation struggle has been characterized as extremely brutal and sexist, while in Afghanistan, the Taliban are using online propaganda to fight their opponents (Pollet, 2021).

A controversial situation also arose when France undertook offensive actions in Africa to counteract Russian disinformation. The French created mechanisms and structures analogous to those in Russia, including a network of false accounts, assuming that they would be able to more effectively limit Moscow's influence over the population of Francophone Africa (Brandt, 2021a). Yet democracies imitating Russia or other active actors in the world of disinformation is counterproductive. It requires submission to authoritarian rules and it legitimizes the disinformation actions of opponents. One may wonder whether Paris, instead of imitating Moscow, should rather appeal to its allies and partners and use political, diplomatic, economic, and cyber tools to increase Russia's costs for such operations. Perhaps a beneficial response would be to initiate a complexity of actions with the EU's participation, including counteracting political corruption and the flow of funds from which influence operations are financed. This in turn would help African states and communities attacked by Russia to create their own instruments for countering disinformation and for supporting free media.

Nevertheless, in France's actions one can recognize and appreciate a strong will to actively oppose foreign influences. Such an approach is also distinguished by the development of military means (including offensive actions) and the adoption of the doctrine of military combat and the fight against disinformation. It seems that, among the countries of the Western community, France, the United States, and the United Kingdom are the most advanced in this aspect of counteracting disinformation.

Based on the research conducted by the authors, a global perspective reveals that only a few countries have made significant efforts to create a comprehensive model for countering disinformation. These countries are:

1. Western countries of global or supra-regional international importance.
2. Countries affected by Russia's disinformation for geopolitical and military reasons, such as Ukraine.
3. Countries with a relatively high income and advanced educational models, which predispose them to quick adjustments. These include the Nordic, and to some extent the Baltic, states.
4. Several states that have taken such steps after experiencing interference with sovereign elections.

It is the task of governments, civil societies, and experts to transform these achievements into a comprehensive, integrated, and sustainable but adaptable system for the entire Western community, encompassing as many countries as possible. Systemic solutions can be distinguished in the countries discussed below (Jeangène, 2021).

1.9 *France*

After the 2017 presidential elections in France, a working group was established consisting of experts from the Ministry of Foreign Affairs and the Ministry of National Defense. The group prepared a 200-page report titled *Information manipulation as a challenge to our democracy* and implemented its recommendations. The report contained 50 recommendations, including that civil society should be on the front line of defense in the adopted model. The report also stated that it was necessary to create administrative structures to detect and counter disinformation, improve the transparency of foreign media registration, and introduce parliamentary hearings on these matters. It also recommended making online platforms responsible for the content they show; strengthening international cooperation within the European Union's East StratCom task force (European External Action Service – EEAS) and the NATO Strategic Communication Center of Excellence in Riga; and intensifying media education and critical thinking curricula for children, teenagers, and adults.

The General Secretariat for National Defense and Security at the Prime Minister's Office and the National Cybersecurity Agency reviewed and analyzed threats, then presented their findings to national stakeholders. Actions were taken in public communication and diplomatic warnings against disinformation were also issued, including at the level of the presidents of France and Russia. In addition, a military doctrine of information operations was adopted, which defines operations of influence using the web as "military activities in the information domain of cyberspace in order to detect, assess, and counterattack, support strategic commands, obtain information or misrepresentation, as a standalone operation or as part of a wider activity" (Kolesnyk, 2021).

Legislators introduced penalties for publishing and disseminating false information in the media, with fines of up to 445,000 euros for intentionally spreading false information that violates public order and up to 135,000 euros for false information that interferes with military discipline, morale, or the nation's war effort. In November 2018, a new regulation was adopted to combat electoral disinformation, which mandated transparency in the dissemination of sponsored information and gave the Radio and Television Council the authority to suspend content from media supervised by foreign countries or related to them, subject to judicial review. The Viginum unit for combating external interference was established in 2021 as part of the Prime Minister's office, with a budget of around 12 million euros and 70 staff.

1.10 *Canada*

While not the main target of disinformation operations, Canada has taken steps to make the electoral processes more secure. In 2018, a national cybersecurity strategy was adopted and the Cybersecurity Center was established. The Ministry of Democratic Institutions prepared a plan to ensure safe elections, aiming to improve civic resilience and the readiness of political parties. The plan also aimed to counteract disinformation and foreign interference by the administration. A task force for security threats was created, where state information, communication, and diplomatic services cooperate. Additionally, a new legal act on elections was passed to protect candidates against disinformation, ban financing campaigns from outside, and ensure transparency of campaigns, including a register of political advertisements on social media platforms. The government also outlined specific actions they expected from internet platforms.

The Ministry of Foreign Affairs now has a digital policy center with two teams. One team is responsible for the rapid response mechanism and the other for digitizing foreign policy and dealing with the interconnections of digital technologies, moderation on social networks, and issues of artificial intelligence and digital education. Canadian Heritage – country's the Ministry of Culture – is responsible for media education of civil society. The Government of Canada also works together with social media platforms. During the 2018 G7 presidency, Ottawa initiated the establishment of a Rapid Information Exchange Mechanism in the event of disinformation attacks and brought about an agreed framework. It has traditionally supported Ukraine's efforts to combat Russian disinformation, including by providing financial support. In 2020, a citizens' commission was established, and broad public consultations were initiated on regulating the digital market and social media platforms in terms of countering disinformation and hate speech.

1.11 *Germany*

Due to its prominent international position, Germany is a country that faces a high degree of risk and threats relating to disinformation, particularly from Russia. To address these concerns, the Federal Union Treaty on the Media was amended, granting authorities in the union lands the right to initiate proceedings against media disinformation. Additionally, the obligation to mark advertising materials more transparently when using bots was introduced. The role of public service media has also been strengthened, requiring greater availability of content from such media on other platforms. The German law of 2017 on improving law enforcement online was adopted in response to increased hate speech and disinformation, and it regulates the procedure for complaints about illegal content. This is defined in the criminal code and establishes time limits for removing manifestly illegal content, with removal required within 24 hours or 7 days depending on the nature of the content.

In preparation for the 2021 parliamentary elections, various measures were taken in Germany to combat disinformation. These included public warnings by the spokesperson of the Ministry of Foreign Affairs and publicly attributing disinformation to perpetrators associated with the Russian authorities. Separate websites were set up to provide information about the elections, and the Central Election Commission conducted a dedicated information campaign to promote the transparency of the election process.

Working groups for hybrid threats were established in various state institutions, and knowledge and expertise were integrated into inter-ministerial teams. A team was created at the Ministry of the Interior with representatives from the Ministry of National Defense and the Ministry of Foreign Affairs. The National Cybersecurity Center developed a new cybersecurity strategy that was later adopted by the government. The Federal Office for Information Security conducted training for politicians, decision-makers, and officials and strengthened cooperation with platforms such as Facebook and Google to detect bots and coordinate inauthentic behavior on the web.

Political parties also implemented their own anti-disinformation programs. The Christian Democrats created a fact-checking page, while the Greens established a fire brigade in the *Netfeuerwerk* network. Self-fact-checking initiatives originated in the mass media, such as the DPA-created Fakt21, which focused on training, education, and cooperation among journalistic circles. Similar programs were also launched in the research community of think tanks, including international think tanks.

1.12 *Sweden*

In Sweden, like in Canada and to an extent France, the government's intention is to stay outside the front lines of the fight against disinformation. In this regard, the government in Stockholm has an easier task because Sweden's society is less polarized than in many other Western countries. Furthermore, the opposition thinks similarly to the government in matters of security policy, which creates fewer opportunities for divisions and therefore for external actors to play on them. Swedish society, which also has a high level of general education, is a resilient community largely unaffected by influence operations. In 2018, both the government and civil society, drawing on the experiences of other countries including the U.S., created a response system. A nationwide media fact-checking platform covering the most important media was established as part of it. Foreign-funded advertising was banned, and relevant teaching materials for pupils and high school students were developed. Based on the experience gained, the Swedish Civil Contingencies Agency (MSB) issued a special guide for public officials and officials dealing with social communication (*Countering Information,* 2019).

MSB, transformed into the Swedish Psychological Defense Agency in 2022, is closely involved with media education in partnership with regions and local authorities as well as social groups and citizens. It cooperates with the private sector, the media, and public relations companies. It also finances research to support subsequent training and education. In addition, it carries out preventive work, including public outreach campaigns.

Sweden works closely with other Nordic countries and the Baltic countries, which look to Sweden for solutions to develop their own prevention systems. The handbook for officials commissioned by the MSB was developed in cooperation with the United Kingdom. Sweden prioritizes the participation of its representatives in international efforts, such as the EU's special task force (EEAS) to combat disinformation and the NATO-affiliated Center of Excellence for Strategic Communication in Riga.

The Swedish system for countering disinformation focuses on key functions, such as coordination, inter-ministerial and international cooperation, education, research, exercises, and strategic communication. Its efforts have proven effective in deterring foreign disinformers.

1.13 *United Kingdom*

London plays a significant role in the international fight against disinformation on account of its experience, global influence and interests, and the worldwide reach of British media and research organizations. British experts have been instrumental in shaping the premises of NATO's communication strategy and are actively involved in its implementation. They have adapted the OASIS (*Objective, Audience, Strategy, Implementation, Scoring*) model of information campaigns for use in NATO, and it is also utilized as a tool to verify the effects of strategic communication efforts carried out by allies. The British have also developed a toolkit for countering disinformation for their public officials, which was developed in partnership with Sweden.

The British approach is also characterized by networking and expertise. There are units in the Office of the Prime Minister and the government chancellery, including teams for communication security, rapid response, and media monitoring. In the Ministry of Foreign Affairs, in addition to the

Russia-focused team, there is an open-source information unit that collects data from research and open intelligence sources. As in Sweden, the British approach distinguishes between two main currents of counteracting: dealing with disinformation in general and targeting specific state disinformers. In connection with the COVID-19 pandemic, a separate interdepartmental coordination team was also created.

1.14 *United States*

The comprehensiveness of the US approach may be debated, but it is showing stronger foundations for countering disinformation in a systematic manner. This was largely spurred by foreign interference in electoral processes, as well as the recent experience of the insurrection on January 6, 2021, when supporters of then-President Donald Trump staged riots at the U.S. Capitol in Washington, resulting in the deaths of six people.

Russia has declared a veritable information war against the United States, the leader of the world's democratic community. The effects of this campaign, as well as American countermeasures, are reflected in official documents, such as the Robert Mueller report on Russia's interference in the 2016 U.S. presidential elections, and numerous reports by research centers, including those of Harvard University, the University of Texas at Austin, and the Massachusetts Institute of Technology. These reports are important sources of conclusions and recommendations for experts, decision-makers, and practitioners in other countries in the Western community.

Despite controversy during Trump's presidency, the United States responded to Russia's disinformation attacks and influence operations with sanctions against the perpetrators. Before the 2018 midterm elections, the American government conducted preventive cyber operations against the Russian troll factory in St. Petersburg. In 2017, as part of a defense package, legislation was passed to counter foreign propaganda. The Departments of State and Defense were mandated to develop a strategy that included assistance to third countries. The Global Engagement Center was established in the Department of State, and it focuses on information technology issues, international cooperation, and the preparation of materials to counter disinformation. The center also works together with U.S. security services.

Senator Amy Klobuchar has introduced a bill in the Senate that would hold companies responsible for allowing misleading information about vaccines and other health issues spread on the internet. The bill proposes to introduce an exception to the current internet law, which has protected companies such as Facebook, Google, and Twitter from legal accusations related to content published on their platforms. The sensitivity of this issue was demonstrated by limiting liability to current threats to public health, such as epidemics or other situations with mass-scale consequences. The proposed act would not apply when the publication appears organically, such as when it was created by a person rather than an algorithm capable of duplicating it many times (Ghaffary & Heilweil, 2021).

California commissioned RAND to diagnose the disinformation problem during the 2020 elections and make recommendations for the future. The research showed that content prepared by Russian-associated perpetrators was considered by Republican-leaning voters to be a product of the Democratic Party, and vice versa. The materials covered public and social affairs, dividing American voters. One of the researchers recommended that authorities publicly inform the public about the perpetrators and the content of such actions during election campaigns via Public Service Announcements, stating, "Russia knows who does not like whom and what causes divisions, and fills the information space with messages that hinder agreements" (Posard et al., 2021). It is also worth noting instances of local initiatives by the governors of some U.S. states, despite raising controversy. For example, Florida has banned removing candidates' accounts in elections for state and local office, while Texas has banned content moderation by social media platforms, claiming they are supposed to be treated like telephone service providers. Florida's bill has been criticized, however, as an anti-democratic method of tackling undemocratic problems.

1.15 *Poland*

In 2018, Poland was ranked by renowned Czech think tank European Values as one of the countries with a higher awareness of threats related to countering disinformation, mainly among state authorities (Víchová & Janda, 2018). From a broader perspective, however, Poland's weaknesses were identified as the selectivity of actions in cyber-security, dispersion of competences at the administrative levels, and neglect in education and support for independent media.

To counter disinformation, dedicated units have been established in the state administration, including the Ministry of Foreign Affairs and the Ministry of National Defense. The MFA operates the EU Rapid Alert System (RAS) and conducts training in this area, including for the top management of central bodies. The National Security Bureau, the Department of National Security at the Chancellery of the Prime Minister, intelligence and counterintelligence services, the Government Center for Security, and the National Broadcasting Council, which monitors political advertisements, also all deal with issues relating to disinformation. The government administration's activities are supported by the Scientific and Academic Computer Network (NASK) – the National Research Institute, which answers to the Ministry of Development.

As regards legal issues, Poland's constitution enshrines the freedom of expression in Article 54, while other laws impose certain obligations on the media and specify the conditions for granting broadcasting licenses. The Penal Code defines offenses related to the dissemination of false information. In January 2021, the Ministry of Justice presented a draft law on freedom of speech that deserves special attention due to its potential importance for the information environment in Poland and the fight against disinformation. The drafted law defines the concept of disinformation as false, manipulated, or misleading information that is for profit or with the purpose of violating the public interest. It also refers to the criminal code regarding offenses infringing on personal rights and sets out special obligations for service providers, including semi-annual reporting if they receive more than 100 complaints per year for content posted on their platforms. The draft establishes the Freedom of Speech Council with wide prerogatives to intervene and impose very high financial penalties for misdemeanors. However, the draft does not refer directly to the threats of disinformation by foreign state or non-state perpetrators (Ministry of Justice of the Republic of Poland, 2021).

Commenting on the project, the Polish Ombudsman, in his formal opinion (Public Information Office of the Ombudsman, 2021) opines:

- It is reasonable to remove or block content on social networking sites; however, the draft does not include definitions as to what hate speech is.
- The definition of illegal content did not include the necessary prohibition of discrimination on the grounds of sex, race, ethnic origin, religion, or sexual orientation. The Ombudsman recommended the use of the definition adopted by the Council of Europe.
- It is justified to protect the right to truthful information and the defense against disinformation, content of a criminal nature, or content violating decency, disseminating or praising violence, suffering and humiliation.
- It is difficult to determine the exact number of people who use social media service providers and their portals due to the nature of online communities, the fact that users can choose to be anonymous, and the existence of automated processes.

The Ombudsman also pointed out that the proposed changes to the Electoral Code were questionable and, above all, they pointed to the risk of restricting freedom of speech through arbitrary interventions by the Freedom of Speech Council. This last remark is especially relevant as the method through which members of the council are selected also raises doubts. If it is not possible to select them by a three-fourths majority in Parliament, they may be elected by a simple majority in the next vote; this stipulation causes concern around its potential to discourage pluralism. The Ombudsman aptly noted that protecting users of internet portals will be more effective within the framework of uniform European standards. It is therefore advisable to postpone the procedure of the project until works in the EU have ended and until the implementation of the Digital Services Act has started.

2 Counteracting Disinformation: a Regional Perspective

Based on the information available, one could consider examining the issue of combating disinformation in Europe from a regional perspective. Northern Europe appears to be the most advanced region in countering disinformation from a political and organizational point of view, and it is characterized by a strong civic, proactive, and integrated approach to the problem. In contrast, the eastern and southern parts of the continent present a mosaic of disjointed, chaotic, or non-existent measures against disinformation. A special case is Ukraine, where after 2014 geopolitics and reality forced the country to deal with unprecedented challenges in the information and cyberspace environment. Additionally, many countries have recently adopted cybersecurity strategies as a response to disinformation threats.

2.1 *Nordic Countries*

The Nordic countries are known for their exemplary resistance to disinformation, including from Russia. Their success can be attributed to their unique societies that are historically shaped by social solidarity, economic strength, high levels of education, and quality media. Finland and Sweden have particularly effective models for countering disinformation, backed by organized education systems that promote critical thinking and creativity at all levels. The Nordic models also employ severe penalties for disinformation. Denmark, for instance, penalizes acts related to media influence and disinformation by foreign states during election campaigns with penalties of up to 12 years imprisonment, although this provision does not cover social media.

2.2 *Baltic States*

Estonia, Latvia and Lithuania are targeted by Russian disinformation due to their geographical location, membership in the EU and NATO, the presence of NATO military forces in their territory, and the existence of significant Russian minorities in Estonia and Latvia. As a result, these countries invest more in their capacity to counter disinformation and host various analytical institutions, including Centers of Excellence affiliated with NATO: cyberdefense (Estonia), strategic communication (Latvia), and energy security (Lithuania). They have adopted the Nordic countries' solutions to countering disinformation, leveraging their regional proximity and high level of cooperation within the broader EU and NATO framework. Among these three countries, Lithuania is most active in countering disinformation through military means, and it is also the most severely attacked due to its more assertive policy toward countries such as Belarus and Ukraine, which Russia perceives as anti-Russian.

Latvia and Estonia have refused to register branches of Russian state media outlets *RT* and *Sputnik*. However, as they have not yet created an appealing alternative program for Russian-speaking minorities, these minorities still largely remain in the pro-Kremlin information space. In 2017, the Latvian Ministry of Culture launched training programs for journalists in investigative journalism, fact-checking, and media education. Estonia has taken similar actions, introducing compulsory media education in secondary schools. In Lithuania, there is a kind of militarization of the media space, treated by the authorities almost as a separate domain of military operations. Military and civilian experts in psychology, social sciences, cyber security, and intelligence monitor the media, analyze it, and react by reporting incidents that could affect state security. The cooperation between government institutions and society enables Lithuanians to effectively mitigate the impact of disinformation and clean up the information space in a concrete way while protecting decision-making processes. This is a phenomenon of massive societal involvement in the informational security of the state by professional and volunteer teams through projects such as Debunk EU and the Elves movement.

2.3 *Visegrad Countries*

The Visegrad Four countries, namely Czechia, Hungary, Poland, and Slovakia, do not have a homogeneous response to Russian activities and methods of influence, including their perception of threats and responses to them. Their credibility and role of public media also differ. Of the four, Poland is targeted the most by Russia due to historical disputes, regional ambitions, and its role in the EU and NATO. While direct pro-Kremlin disinformation is ignored in Poland, extremely politicized messages are used in public media that exploit phobias, extremisms, conservatism, parochial religiosity, and specific historical examples. These messages deepen social divisions and are conducive to information manipulation by external entities.

In recent years, there has been a stronger desire among authorities in Slovakia to pursue a more proactive information policy. This is taking shape in the form of adopting numerous new documents of a doctrinal or strategic nature. These documents include those on hybrid threats and disinformation, as well as the establishment of new cells in the government's Situation Center and the Government Center for Security Analysis. In 2021, a new security strategy was adopted with provisions on countering disinformation. Additionally, Slovakia's participation in the NATO-sponsored information campaign resulted in a rise in public support for the Alliance in 2020 and 2021 by several points, reaching over 60%.

In Hungary, the close business ties between the political and economic elites and Russia have resulted in Moscow's messaging infiltrating the Hungarian information space through local mainstream media. Furthermore, the media consistently pursues a policy of discrediting the European Union and Western circles. The approach to Russian propaganda and disinformation in Hungary differs significantly from that of other Visegrad Group countries. This divide has been further heightened by disagreements over sanctions against Russia and support for Ukraine after the aggression in February 2022.

Czechia was one of the first countries in the entire EU, and the first in the region, to establish a dedicated unit in the Ministry of Interior and an inter-ministerial structure to counteract disinformation. Taking inspiration from the Baltic states, Czechia employs "elves" who track accounts and online platforms and monitor other related activities, such as campaigns to spread disinformation about COVID-19 (Zamecnik, 2021). The Czech Demagogue has become a source of inspiration for Polish fact-checkers. In 2021, the country adopted a new Strategy for Counteracting Hybrid Threats.

2.4 *Southeast Europe*

In the countries of this region, aside from Ukraine and Romania, the sense of being threatened by foreign disinformation is limited. The COVID-19 pandemic and the related increase in general online disinformation, however, have led to reactions from local governments. Additionally, due to the presence of NATO structures and allied troops on its territory, Romania often experiences Russian disinformation campaigns.

Ukraine is a unique case, in terms of both Russian disinformation campaigns directed against it and its experience and activity in countering them, particularly in the realm of a defensive war following Russia's aggression. In partnership with the European Union and NATO, Ukraine runs many projects aimed at strengthening the media and combating disinformation. The Ukrainian authorities, particularly since the presidency of Volodymyr Zelensky, have undertaken the daunting task of limiting the presence of pro-Kremlin national media in the Ukrainian information space. This effort included revoking many licenses and blocking access to the most popular Russian-language social networking site, VKontakte. The country has adopted doctrinal and strategic foundations for countering disinformation and established institutions and teams with coordinating functions in the Office of the President of Ukraine, the Security and National Defense Council, and the Ministry of Culture and Information Policy. These teams are actively involved in media education, particularly for schools in eastern Ukraine.

The role of independent Ukrainian media and civil society cannot be overstated, as the supra-regional project called StopFake, including the one existing in Poland, was designed at the Mogilev Academy in Kiev. Civic circles have initiated projects to monitor the image of the state abroad, including one by the Ukrainian diaspora (Havelock & Veliseyeu, 2021). Ukraine has shown that it can effectively defend itself against Russian disinformation during the war and has provided many examples of attempts to reach recipients in Russia itself using popular social channels on Telegram. Ukraine's efforts in combating disinformation undoubtedly provide a uniquely interesting research field, with conclusions that Western countries could use to improve their own tools for combating disinformation.

2.5 *Southwest Europe*

In response to Russia's interference in internal affairs related to Catalonia, Spain has taken the most concrete steps against disinformation among the countries in the region. Its government has adopted relevant action plans and established structures to fight disinformation.

Italy, where the level of pro-Russian sympathies is among the highest in Western Europe and where the political mainstream parties cooperate with partners from Russia, does not seem to be particularly preoccupied with the problem of disinformation, at least at the governmental level. It is, however, the subject of research and prevention of Italian expert and journalistic circles. In 2021, an interdisciplinary national center for countering disinformation was inaugurated at one of the universities in Rome as part of the pan-European EU-supported network, European Digital Monitoring Observatory (EDMO). This network specializes in fact-checking and brings together many organizations from the countries of the region, including Greece, Portugal and Spain. It is unclear, however, whether Russian disinformation is a serious problem on the agenda of the governments of the former two and Cyprus.

2.6 *Benelux*

In the Benelux countries, Russian disinformation is primarily addressed by civil society and the media rather than the government. In 2018, the Belgian government made its first significant attempt to develop a response to disinformation by allocating funds to financially support non-governmental organizations. In the Netherlands, steps have been taken to expose Russian disinformation, including relating to the shooting down of a Malaysian airline plane with Dutch citizens in 2014, as well as an attempted hacking attack on the headquarters of the Organization for the Prohibition of Chemical Weapons in The Hague in 2018. The Dutch organization DROG has had success in counteracting international disinformation through training programs, NATO officer training, and simulation games. The Dutch model emphasizes the credibility of traditional media, which is trusted by the vast majority of society, and public television runs media education projects. Leiden University is a strong center for fact-checking.

2.7 *Global Considerations*

On a global scale, Australia has developed one of the most effective models for countering disinformation. This model, however, has been primarily influenced by the threat of Chinese operations rather than Russian ones. According to the GDI study *Disrupting Disinformation: A Comparative Analysis of Regulatory Frameworks for Countering Disinformation* published in 2021, Australia is the only country that meets the basic systemic criteria for resisting disinformation. This includes combatting hatred, ensuring transparency and effective organization in elections and the functioning of government institutions and task forces, and imposing sanctions on media that violate regulations.

Increased internet control measures are often a response to local tensions and rivalries, as seen in the case of India and Pakistan. In order to limit tragedies such as the scourge of public lynching, often a result of false information circulated on the web, India introduced restrictive regulations. Such disinformation is largely spread through WhatsApp, a particularly popular platform in India.

In Indonesia and Turkey, the authorities' control measures aim less to protect citizens' interests and more to protect the interests of the ruling parties. Even nominally independent fact-checking organizations, as is the case in Turkey, are focused on monitoring the country's image abroad. In Egypt, individuals with over 5,000 followers on their social media accounts are required to register as media organizations.

After studying the analyzed cases of individual countries, three basic models of the approach of state authorities to regulating the internet can be identified. They are:

- The democratic model, focused on the needs of citizens including their protection against internal and external disinformation.
- The authoritarian model, dominated by the interests of the ruling class and, to some extent, by concern for the safety of citizens, like in India and Turkey.
- The dictatorial model, which ignores the needs of citizens and focuses on full control of the internet, such as in China and Russia.

Most experts agree that the optimal approach for states to tackle disinformation is through a holistic, multi-sectoral, and supra-ministerial approach. Some countries have already adopted this approach, but partial solutions are still dominant. Additionally, many countries tend to focus on combating disinformation during election campaigns and elections, whereas effective countermeasures should be continuous. Another issue is also how to combine the fight against disinformation by foreign states and actors with disinformation carried out by national organizations like political parties. For instance, the Swedish institution tasked with countering foreign disinformation cannot act against its own citizens who spread disinformation for domestic reasons. Similar constraints apply in many other democracies. Clear guidelines are therefore needed on how to differentiate between foreign and domestic disinformation, as well as on how to respond when governments themselves engage in disinformation campaigns targeting their own citizens.

Based on the reviewed national and regional approaches, several elements of an overall systemic structure for counteracting disinformation at the national level have emerged. No country has implemented a complete system, however, although Australia, the Nordic countries, France, and Germany are the closest to achieving this. A complete system should be based on doctrine and strategy and take into account the following considerations:

- Using legal instruments and executive acts in halting disinformation.
- Creating specialized units in the administration structure.
- Establishing structures for protecting elections.
- Demonstrating the ability to analyze the information environment, identify threats, and adapt to them while recognizing one's own weaknesses and groups susceptible to disinformation in the dominant narrative.
- Increasing social resilience and strengthening the credibility of public institutions and mass media, media education, digital education, and critical thinking. This can be done together with journalists and the media.
- Participating in active strategic communication.
- Conducting research that will identify training needs, including for policymakers, politicians, and journalists.
- Working at the local level and with social groups.

– Cooperating with media and platforms.
– Engaging in international cooperation.

3 The EU, NATO, and the UN: Combating Disinformation

The European Union has taken the most comprehensive action in supporting member states in the fight against disinformation, both in general and specifically against disinformation originating from Russia. This began with landmark decisions made in 2015 following the illegal annexation of Crimea and Russia's aggression against Ukraine. These decisions led to the establishment of a specialized task force called East Strat-Com within the European External Action Service (EEAS). Many practical projects and steps have been taken as part of the European Action Plan on Democracy, resulting in new experiences and insights. The Code of Conduct for Combating Disinformation (Killeen, 2021b) was agreed upon as the next stage, and lessons from its implementation were considered in the enactment of the Digital Services Act (DSA). The act imposes many legal obligations, both voluntary and involuntary, on large platforms such as Facebook, YouTube, and Twitter. These include an obligation to cooperate with independent researchers and to allow them to access and participate in complaint and appeal procedures regarding content moderation, dispute resolution, and access to platform archives.

At the social level, the regulation provides for consultations with civil society organizations, as well as the introduction of institutions for trusted whistleblowers to report suspected violations. The act also establishes a European Digital Services Council (DSA) and an advisory body of national coordinators responsible for implementing legislation at the national level. The DSA defines the responsibility of service providers, their obligations, and rules for handling complaints, including out-of-court dispute resolution. It imposes additional obligations on very large internet platforms, those whose services are used by at least 45 million recipients per month, to assess systemic risks resulting from their services, indicate measures to reduce these risks, undergo independent audits, and have conditional algorithmic recommendations and additional transparency for advertisements (*The Digital*, 2021).

NATO relies heavily on strategic communication and media operations to analyze and counter disinformation and propaganda directed at the Alliance's values, goals, policies, activities, and operations. Rather than denying already circulated news, NATO believes that preventing disinformation is more effective. The Alliance's approach is therefore preventive in nature, with a focus on countering opponents' goals through campaigns that support NATO's role and mission in member and partner states' societies. The Alliance's research and analytical activities are supported by the Center of Excellence for Strategic Communications in Riga and the NATO Defense College in Rome.

In its efforts to combat disinformation, the Alliance has the closest cooperation with the European Union, as well as with the United Nations, the G7, and partner countries. During the COVID-19 pandemic, NATO's steps to counter disinformation demonstrated its ability to sustain its operations, continue its missions and activities, and remain prepared despite the pandemic. This helped ensure that the global health crisis did not escalate into an international security crisis (NATO, 2021).

In 2019, NATO developed a structured package of measures to combat disinformation, which was updated the following year to address hostile disinformation related to COVID-19. In 2021, the Alliance created a toolbox with a two-pronged response model: "understand and act" on one prong and "coordinate" on the other. Its purpose is to provide Allies with instruments to assess hostile information activities and disinformation, and to help identify possible courses of action. Experts from the NATO International Secretariat also organize regular briefings on Russian and other disinformation activities in various Alliance committees, including the Civil Emergency Planning Committee. Additionally, in 2022, the Resilience Committee was established with a mandate to counter disinformation.

As part of the anti-hybrid strategy, Rapid Reaction Teams were put at the disposal of Member States with the participation of strategic communication and counter-disinformation experts. In 2020, NATO further adapted its approach to combating disinformation by increasing support for projects aimed at strengthening social resilience to disinformation in member states. Additional funds were directed towards non-governmental and expert organizations for this purpose.

The role and importance of other international organizations and institutions in countering disinformation are limited compared to NATO and the European Union. Nevertheless, the UN system and the Council of Europe aim to encourage member states to step up their efforts to counter disinformation through actions for media education. These organizations, together with the OSCE, have adopted joint declarations on disinformation threats and media freedom. Since 2018, the OECD has been including the results of media education in PISA surveys. In 2022, the organization launched consultations on the draft *Principles of Good Practices of Response in Public Communication to Disinformation* to support member state governments in improving media and information ecosystems and creating a space for the exchange and dissemination of information that builds social resilience to online and offline false narratives, thus strengthening democracy (*Public consultation*, 2022).

International trade agreements cannot prevent cross-border disinformation, but they can limit its scale. One proposed solution is to create a model for such agreements within the United Nations, specifically the United Nations Commission for the Law on International Trade. This model would define cross-border disinformation and determine how to attribute it to its perpetrators, as well as prohibit companies from producing and spreading disinformation content abroad. These agreements would also impose obligations to introduce such regulations in domestic legislation (Aaronson, 2021).

The United Nations has established an Information and Democracy program, modeled after the UN process of combating climate change and led by the International Observatory on Information and Democracy. States and civic organizations have also signed the associated Partnership for Information and Democracy, and a global Civic Coalition has been established through the website *information.democracy.org* for the same purpose. The UN agenda also includes raising awareness of the damage caused by global disinformation during crisis situations such as the coronavirus pandemic.

4 Recommendations for Strengthening the Activities of International Organizations: from Practice to Strategy

Organizations that bring together democratic states have implemented various measures to protect themselves against disinformation, but their efforts still appear too limited and defensive. Even the North Atlantic Alliance, which has adapted its approach to disinformation by implementing strategic communication and anti-hybrid strategies, has yet to develop a comprehensive strategy for countering disinformation. Given the increasing threats related to disinformation, measures to counter it should be given higher political priority. This is happening in the EU, as evidenced by the political documents and legislative work mentioned above, as well as by assertive public communication at the political level and Member States' positions on Russian actions.

A new strategic concept was adopted at NATO's Madrid summit in July 2022, where countering disinformation was recognized as a critical issue. The summit aimed to encourage member states to make more explicit commitments to combating disinformation and to hold them accountable for fulfilling those commitments. It also noted that NATO could enhance the mandate of existing groups dedicated to countering disinformation and strategic communication to facilitate coordination among national efforts, encourage information sharing about threats, and promote effective response measures.

The West's coherent response and resilience building both nationally and internationally should focus on: (1) strengthening social resilience to disinformation; (2) extending preventive and offensive activities; and (3) bridging the differences in practical approaches to disinformation in individual Western countries.

We recommend the following measures.

4.1 *Strengthening Resilience*

- Emphasize broader education regarding democratic values, improving electoral standards, and monitoring campaigns in terms of transparency, fairness, and funding.
- Enforce a media policy aimed at strengthening trust in the media, with stronger and better financed support for media independence and standards, investigative journalism, and fact-checking.
- Take assertive actions toward social media providers. After the new Digital Services Act is adopted at the level of the European Union, it is necessary to support its implementation for the Member States at the national level.
- Conduct information campaigns and more active teaching in schools about the organization's security policy in NATO countries.
- The Alliance should make better use of the network of embassies acting as points of contact to counter disinformation about NATO in partner countries.

4.2 *Preventive and Offensive Actions*

Western countries should more proactively raise the costs for disinformers by publicly exposing their activities and imposing sanctions. Specifically, they should:

- Identify Russian propaganda media operating abroad and harmonize the decision-making standards of regulatory authorities regarding this media. They can also standardize the procedures for punishing media entities for disinformation, including through a suspension of their activities.
- Expose disinformation and influence operations, including disavowing them at high levels through government statements and reports from special services. Such reports should be made publicly available in all NATO and EU member states.
- Employ the proactive use of alert systems within the EU, NATO, and G7.

European states should also take bolder decisions on sanctions against the employees of Russian and Belarusian propaganda institutions.

Within the EU and NATO, and perhaps also in cooperation with the OSCE, a model and practice of pre-bunking activities should be developed as part of pre-election missions in the member states of these communities. As a result, assessments should be developed and partially made public regarding pre-election threats, including cyber threats, and related countermeasures.

National plans presented to NATO allies and accounted for in the annual planning cycle should also be considered as measures to anticipate and pre-bunk disinformation.

4.3 *Bridging the Gap in Resistance to Disinformation*

This group of activities should include:

- Assigning groups, or a joint group, existing within the EU and NATO, to coordinate the exchange of information and experience on good practices on countering disinformation.
- Developing a toolkit for countering disinformation while using existing models in places like the United Kingdom and Sweden.
- Supporting special projects for the Balkan states from the EU, NATO, and the U.S. using the existing networks of delegations, representations and embassies, and resources and funds. This could involve creating a separate program aimed at civil society and the media.
- Creating programs to support local initiatives in the EU and NATO partner countries, where necessary and possible.

5 Conclusion

The main aim of the conclusion is to present a range of existing systemic, doctrinal, institutional, and operational elements of countering disinformation, incorporating the governmental, social, and individual levels. In addition to national frameworks, international frameworks are also established. The effectiveness of many activities is challenging to measure, and their efficacy is often confirmed by experience rather than research. Nonetheless, the examples discussed above demonstrate their effectiveness.

In the United States, a comprehensive analysis was conducted in response to the 2016 election interference. The government released information on Russian disinformation and implemented coordinated measures to protect cyberspace resources during subsequent elections. Authorities proactively communicated about cybersecurity measures. Social media platforms such as Facebook, Twitter, and YouTube marked election-related content as requiring verification and critical evaluation. Accounts that violated their regulations were suspended. The Partnership for Electoral Integrity was established, and a non-partisan coalition of disinformation researchers identified, tracked, and responded to disinformation. Its main goals were: (1) identifying disinformation before it proliferated; (2) sharing clear, accurate messages about detected disinformation activities; and (3) increasing transparency in the information space.

In connection with the parliamentary elections in 2021 in Germany, the authorities decided to take unprecedented public interventions against Russian disinformation. They took unmasking actions, including warnings at the highest level of the state, attributing perpetration to entities related to the Russian authorities. Hybrid threat teams in various state institutions integrated knowledge and inter-ministerial expertise and increased the efficacy of these efforts.

In Moldova, although a lot of work is still ahead, advancements have been made by legislative decisions and the establishment of governmental institutions to counter disinformation. Media education and the activities of non-governmental organizations contributed to the success of a democratic candidate in the presidential election in 2021. This was a sign that disinformation campaigns became less effective when they were fought more forcefully and when voters consumed more information critically and selectively.

Meanwhile, Russia's invasion of Ukraine in February 2022 created a completely new situation in the fight against Russian information manipulation. The war defied the Kremlin's earlier propaganda about its intentions toward Ukraine as people could see on their smartphones, computers, and television screens the brutal destruction and bodies of those killed in Kiev and its suburbs, in Kharkiv, and in Mariupol, the "Ukrainian Aleppo". The criminal terrorist actions taken by Russia against civilians, including women and children, the deaths of thousands, and the exodus of several million Ukrainians, all brutally exposed the cynicism and real plans of Vladimir Putin and the hypocritical state machine behind him. This propagandistic Waterloo (at the time of writing these words, we do not know the outcome of the war yet) may suggest that the pre-war hybrid actions, including disinformation, were not effective against the West. However, this thesis is not yet proven.

As the rules of peace are replaced by the laws of war, the democratic world rallies behind the victim and condemns the aggressor. However, some argue that if the West had taken decisive sanctions against Russia's propaganda and disinformation apparatus earlier, it could have prevented the war altogether.

This would have demonstrated the West's determination and unity, as well as reinforced its societies' resilience against falsehoods, manipulation, and political corruption from Russia. It would also have signaled a different approach toward Russia than in previous years. During the early stages of information warfare, the Russians were highly effective, while the West was caught unprepared.

Though there is no way of knowing if the war could have been prevented, the Western world stood united in defending the victim of such a brutal aggression and in the face of the attack on the foundations of the international order. Victory over disinformation and authoritarianism is difficult to achieve, however. It remains to be seen when and how Russia will emerge from this war in the long term, and in what direction Russian society will go. It is uncertain if this will mark a bloody farewell to imperial ambitions.

On a global scale, China will undoubtedly learn from the lessons of what happened during the war. The conflict exacerbated existing problems and revealed potential effective solutions and procedures. What may have seemed complicated and requiring difficult arrangements for all members of the European Union became simple and immediately implementable in the face of the war in Europe. The "anti-war" information campaign also broke the monopoly of states, traditional media, and specialized non-governmental organizations in combating disinformation in a spectacular way. The activity of ordinary internet users and groups, such as Anonymous, was astonishing as no one suspected their willingness to join the fight against disinformation on such a large scale. The war also highlighted the importance of leadership, exemplified by Ukrainian President Volodymyr Zelensky. It demonstrated the absolute domination of social media in today's information environment, its strength, and its double-sidedness, as exemplified by the channels used by Russian authorities on the Telegram platform.

The sanctions imposed after the onset of the war met with the expected countermeasures by the Russian regime, including blocking access to Western social and traditional media, and the closure of the last independent editorial offices in Russia. The challenge has become less about defending against Russian disinformation in the West and more about reaching the indoctrinated Russian society.

The Russian disinformation wall created by Putin is not impenetrable. Millions of Russians have installed VPNs to bypass censorship, with 6.4 million Russians installing them in the first three weeks of the war from Apple and Google applications, compared to 230,000 in the three weeks prior. They also use Tor technology to create portals and networks and receive tens of millions of pieces of information about the war via text messages, emails, and online advertisements. Traditional media also play a role in reaching Russian audiences, with newspapers in Nordic countries, Poland, and Germany publishing materials about the war in Russian or Ukrainian. Most Russians are still heavily influenced by the regime's propaganda, however, with the most important disinformation battleground being large cities, primarily Moscow, and younger generations. Putin has closed the last independent media outlets in response to this.

The extraordinary NATO summit on March 24, 2022, meanwhile, decided that the Alliance will continue to oppose Russia's lies about its war in Ukraine and expose fabricated narratives, operations, and provocations. It also resolved to strengthen the resilience of member states' critical infrastructure and societies to Russian influence, including bolstering cyber-defense capabilities and response to disinformation. Additionally, NATO called on China to stop amplifying false narratives from the Kremlin, particularly about the war in Ukraine and NATO (NATO, 2022). Meanwhile, the European Union adopted the Strategic Compass for the future of international security, which addresses foreign influence and manipulation of information (*A Strategic Compass for the EU*, 2022).

Only time will tell how durable the determination of the Alliance, the European Union, and the West will prove to be in the struggle against disinformation in international politics. It already turns out, however, that such a fight does not have to be unrealistic. On the contrary – the free world is well-placed to win, perhaps especially, in circumstances of war.

Summary

1. What is resilience to disinformation?

Resilience to disinformation refers to the capacity to withstand, confront, and effectively address challenges within the information environment at the individual, societal, and governmental levels, including both civil and military domains. It encompasses the ability to recognize and solve problems, evaluate circumstances and reactions, and take appropriate action in response to false, manipulated, or inaccurately presented information that is systematically and persistently disseminated with the intention of causing harm to individuals or groups.

2. Who is responsible for countering disinformation?

Governments, the media, and civil society, as well as individual participants of processes taking place in the information space should all be involved in countering disinformation. Due to the global nature of the phenomenon, some countermeasures must also be international. It is necessary to integrate efforts at every level of counteracting disinformation, and in particular to create media education programs and legal regulations concerning media, freedom of speech, media market and digital services. Active engagement from social groups and professional communities, particularly journalists, academics, and teachers, is also essential in the fight against disinformation.

3. What is media education and what are its goals?

Media education involves developing an understanding, knowledge, and skills that empower citizens to use the media effectively and safely while also thinking critically to make informed judgments, analyze complex situations, and distinguish between opinions and facts. The ultimate goal of media education is to promote media literacy and pro-social and civic behavior. Essential abilities include accessing a variety of sources to find, use, and share information; evaluating and analyzing the quality, truthfulness, and credibility of different viewpoints; creating content and expressing oneself confidently while being aware of the intended audience and purpose; behaving in a socially responsible and ethical manner that aligns with one's own beliefs and the goals of communication; and engaging socially and civically through the media to achieve political self-realization based on democratic values and attitudes.

 | DOI:10.1163/9789004715769_017

4. What does preventing the spread of disinformation at the individual level look like?

To effectively counter disinformation, particularly on social media, it is crucial to internalize certain rules of conduct, including verifying the credibility of sources, platforms, internet addresses, and contact details. This involves checking the credibility of the author and their previous publications, posts, and profile history, as well as scrutinizing the integrity of the text to ensure that it does not contradict itself or other sources and that it is not intended as a joke or satire. Checking the publication date and chronology, as well as paying attention to photos or images for signs of manipulation, is also essential. Additionally, information should be cross-checked in other sources, including with experts, and attention should be paid to the titles and the validity of quotes and expert sources cited. It is also beneficial for individuals to participate in media education and fact-checking initiatives.

5. What is the operating model of social media platforms?

The nature of technology and digital space creates particular challenges related to the spread of disinformation and propaganda, particularly on platforms. These platforms combine four key phenomena in the information space: the aggregation of data about users and their behavior; the algorithmic management of data using computer programs with advanced data processing capacity; the anonymity of aggregation, management, and dissemination of information; and the automation of content publication and user interaction.

Data emission, which is a product of users' activity on the internet, allows platform managers to collect information about users, including their political or election preferences, and to adjust paid communications or advertisements, not only for commercial but also political purposes.

6. What is content moderation?

The primary means of eliminating unwanted content on social media platforms is through moderation. Moderation involves managing and administering the behavior of social media users and involves interference, or lack thereof, in the content generated by users on platforms such as Facebook, Instagram, Twitter, and others, while adhering to their terms and conditions. Moderation is carried out through certain rules and guidelines for posting content, which also restrict and prohibit the posting of unacceptable and inappropriate content. It can occur before publication (preventive moderation) or after it; be algorithmic (automatic) or human-driven. Most platforms use mixed models of moderation. Additionally, moderation may result in a temporary exclusion, such as account suspension, or a permanent exclusion, such as account deletion, from the platform's community.

7. What is the role of researchers and the media in countering disinformation?

Civil society, including research communities, plays a key role in countering disinformation. It is an environment of exerting pressure, especially on governments, to bring about the desired changes. It helps to recognize disinformation and understand its essence. It provides expertise, advice, and a training base for public service employees. It educates users and actors in the information space.

Journalists are on the front lines of the fight against disinformation. Free media, science and education based on truth and freedom are the foundations of democracy. Because of their role in society, journalists should, above all, avoid duplicating disinformation. One of the main weapons in the fight against disinformation is fact-checking, conducted by the media, and its aim is to improve the quality of public debate and to verify the statements of politicians, officials, or other influential people who find their way into the public space. It is also desirable to broadly include the media community in media education programs in schools and local communities.

8. What is the role of state authorities in countering disinformation?

The most effective national solutions for countering disinformation involve active participation from social and professional groups, supported by the state. The optimal approach is a holistic, multi-sectoral, and multi-departmental one, known as the "whole-of-government approach." This system should be based on a doctrine and strategy, taking into account adequate legal instruments and executive acts, the creation of specialized units within the administration, structures to protect elections, and active strategic communication.

The system must be adaptable to changes in the information environment and provide conditions for developing social resilience to disinformation, encouraging cooperation between the government, media, civic organizations, and corporations that control key social media platforms. Governments also have a responsibility to engage in effective international cooperation in the fight against disinformation.

9. What is the role of international organizations in countering disinformation?

Given the diversity and global reach of actors, goals, and methods involved in manipulating content by both domestic and foreign perpetrators, effective solutions to counter disinformation can only be developed through collaborative efforts at both the national and international levels. Organizations comprising democratic states have introduced many political and practical measures to safeguard their member states, but their nature still appears too defensive. Combating disinformation should therefore be given higher political priority. A coherent Western response, building resilience both nationally and internationally, should focus on: (1) further strengthening social resilience to disinformation; (2) offensive activities as much as defensive ones; and (3) leveling the differences in the systemic and practical approach to disinformation in individual Western countries.

10. What is the Digital Services Act (DSA)?

The DSA is a groundbreaking piece of legislation that will fundamentally change the information environment in the European Union, member states, and partner states. Its impact will be felt on a global scale as well. The legislation imposes a number of legal obligations on online platforms, including mandatory ones, unlike the Code of Conduct for Counteracting Disinformation. These obligations include the requirement to cooperate with independent researchers, allowing them access to participate in complaint and appeal procedures related to content moderation and dispute resolution, as well as access to archives and data concerning portal usage rules.

At the social level, it involves consultations with civil society organizations and the introduction of trusted entities to signal potentially criminal behaviors. The act also establishes a European Digital Services Council and its advisory body of national digital service coordinators responsible for implementing rules at the national level. The DSA defines service providers' responsibilities and obligations, including organizational and reporting requirements, and rules for considering complaints, including out-of-court dispute resolution. It imposes additional obligations on very large internet platforms used by 45 million recipients per month. These obligations include assessments of systemic risks, measures to reduce these risks, independent audits, algorithmic recommendation conditionality, and transparency of advertisements.

Bibliography

A Strategic Compass for the EU. (2022). European External Action Service. https://eeas.europa.eu/headquarters/headquarters-homepage_en/106337/A%20Strategic%20Compass%20for%20the%20EU

AAP-6 NATO Glossary of Terms and Definitions. (2021). NATO Standardization Office

Aaronson, S. (2021). Could Trade Agreements Help Address the Wicked Problem of Cross-Border Disinformation? *CIGI Papers*, no. 255. https://www.cigionline.org/publications/could-trade-agreements-help-address-the-wicked-problem-of-cross-border-disinformation/

Adamowski, J. W. (Ed.). (2008). *Wybrane zagraniczne systemy medialne* [*Selected Foreign Media Systems*]. Wydawnictwa Akademickie i Profesjonalne.

Afeltowicz, Ł., & Pietrowicz, K. (2013). *Maszyny społeczne. Współczesna inżynieria społeczna i innowacje socjotechniczne* [*Social Machines: Contemporary Social Engineering and Sociotechnical Innovations*]. Wydawnictwo Naukowe PWN.

Aleksandrowicz, T. (2016). *Podstawy walki informacyjnej* [*Basics of Information Warfare*]. Editions Spotkania.

Aleksandrowicz, T. (2018). Bezpieczeństwo informacyjne państwa [Information Security of the State]. *Studia Politologiczne*, 49, 33–50.

Allcott, H., & Gentzkow, M. (2017). Social media and fake news in the 2016 election. *Journal of Economic Perspectives*, 31(2), 211–236. DOI: 10.1257/jep.31.2.211.

Andrew, C., & Gordijewski, O. (1990). *KGB: The Inside Story*. Hodder & Stoughton Ltd.

Aneja, A., & Ifraimova, S. (2021). How to spot a Russian troll. *Time*. https://time.com/5274785/how-to-spot-a-russian-troll/?xid=tcoshare

Applebaum, A. (2021). *The Twilight of Democracy*. Penguin Random House.

Applebaum, A., Pomerantsev, P., & Smith, M. (2017). *"Make Germany Great Again": Kremlin, Alt-Right and International Influences in the 2017 German Elections*. Institute for Strategic Dialogue. https://www.isdglobal.org/isd-publications/make-germany-great-again-kremlin-alt-right-and-international-influences-in-the-2017-german-elections/

Aristotle. (2002). *Nicomachean Ethics*. Oxford University Press.

Aro, J. (2022). *Putin's Trolls: On the Frontlines of Russia's Information War Against the World*. Ig Publishing.

Aronhime, L., et al. (2021, May 20). Countering cognitive warfare: Awareness and resilience. *NATO Review*. https://www.nato.int/docu/review/articles/2021/05/20/countering-cognitive-warfare-awareness-and-resilience/index.html

Asmus, R. D. (2010). *A Little War That Shook the World: Georgia, Russia, and the Future of the West*. St. Martin's Press.

Astor, M. (2019, February 26). How the politically unthinkable can become mainstream. *New York Times*. https://www.nytimes.com/2019/02/26/us/politics/overton-window-democrats.html

Auerbach, J., & Castronovo, R. (Eds.). (2013). *The Oxford Handbook of Propaganda Studies*. Oxford University Press.

Babik, W. (2014). *Ekologia informacji [Ecology of information]*. Wydawnictwo Uniwersytetu Jagiellońskiego, Kraków.

Baker, G., Faxon-Mills, S., Huguet, H., Pane, J., & Hamilton, L. (2021). Approaches and Obstacles to Promoting Media Education in U.S. Schools. RAND Corporation. https://www.rand.org/pubs/research_reports/RRA112-19.html

Bal, M. (2009). *Narratology. Introduction to the Theory of Narrative*. University of Toronto Press.

Baldwin, D. A. (1997). The concept of security, *Review of International Studies*, 23(1), 5–26. https://doi.org/10.1017/S0260210597000053

Balzacq, T. (2005). The three faces of securitization: Political agency, audience and context. *European Journal of International Relations*, 11(2), 171–201. https://doi.org/10.1177/1354066105050

Band, G. (2021, July 22). *A government practitioner's guide to countering online foreign covert influence*. Lawfare. https://www.lawfareblog.com/government-practitioners-guide-countering-online-foreign-covert-influence

Baraniuk, K. (2017). *Działalność służb wywiadowczych Federacji Rosyjskiej w świetle raportów służb specjalnych wybranych państw Unii Europejskiej [The activity of the Russian Federation intelligence services in the light of the reports of the secret services of selected countries of the European Union]*. Wydawnictwo Adam Marszałek.

Barbashin, A., & Graef, A. (2019). *Thinking Foreign Policy in Russia: Think Tanks and Grand Narratives*. Atlantic Council. https://www.atlanticcouncil.org/in-depth-research-reports/report/thinking-foreign-policy-in-russia-think-tanks-and-grand-narratives/

Barclay, D.A. (2018). *Fake News, Propaganda, and Plain Old Lies: How to Find a Trustworthy Information in the Digital Age*. Rowman & Littlefield.

Barclay, L. (2021, October). *Facebook banned me for life because I help people useitless*.Slate.https://slate.com/technology/2021/10/facebook-unfollow-everything-cease-desist.html?utm_medium=social&utm_campaign=traffic&utm_source=article&utm_content=twitter_share

Barnes, J.E., & Sanger, D.E. (2020, July 28). Russian intelligence agencies push disinformation on pandemic. *New York Times*. https://www.nytimes.com/2020/07/28/us/politics/russia-disinformation-coronavirus.html (accessed: 8.02.2022).

Basaj, K. (2018). Dezinformacja, czyli sztuka manipulacji [Disinformation: the art of manipulation]. *Biuletyn Kwartalny RCB*, 25, 14–17. https://www.rcb.gov.pl/dezinformacja-czyli-sztuka-manipulacji/

Bassat, O.B., & Cohen, I. (2019, September 24). *Mapping the Connections Inside Russia's APT Ecosystem*. Intezer. https://www.intezer.com/blog/malware-analysis/russian-apt-ecosystem/

Bateman, J. (2021, September 13). Germany braces for election disinformation. A growing conspiracy movement is likely to spread false narratives about the results, with echoes of Trump. *Foreign Policy*. https://foreignpolicy.com/2021/09/13/germany-election-disinformation-social-media/

Bateman, J., Hickok, E., Shapiro, J., Courchesne, L., & Ilhardt, J. (2021). *Measuring the Efficacy of Influence Operations Countermeasures: Key Findings and Gaps From Empirical Research*. Carnegie Endowment. https://carnegieendowment.org/2021/09/21/measuring-efficacy-of-influence-operations-countermeasures-key-findings-and-gaps-from-empirical-research-pub-85389

Bay, S., & Fredheim, R. (2019). *Falling Behind: How Social Media Companies are Failing to Combat Inauthentic Behaviour Online*. Riga: NATO Strategic Communications Centre of Excellence. https://stratcomcoe.org/publications/how-social-media-companies-are-failing-to-combat-inauthentic-behaviour-online/33

Bayer, J. (2021, October 13). *Policies and measures to counter disinformation in Germany: The power of informational communities*. Heinrich Boll Stiftung. https://eu.boell.org/en/2021/10/13/policies-and-measures-counter-disinformation-germany-power-informational-communities

Bayer, J., Bitiukova, N., Bard, P., Szakacs, J., Alemano, A., & Uszkiewicz, E. (2019). *Disinformation and propaganda – impact on the functioning of the rule of law in the EU and its Member States. Study for the Policy Department C: Citizens' Rights and Constitutional Affairs*. European Parliament. https://www.europarl.europa.eu/RegData/etudes/STUD/2019/608864/IPOL_STU(2019)608864_EN.pdf

Bayer, J., Holznagel, B., Lubianiec, K., Pintea, A., Schmitt, J., Szakacs, J., & Uszkiewicz, E. (2021, April). *Disinformation and propaganda: Impact on the functioning of the rule of law and democratic processes in the EU and its Member States*. Policy Department for External Relations Directorate General for External Policies of the European Union. https://www.europarl.europa.eu/RegData/etudes/STUD/2021/653633/EXPO_STU(2021)653633_EN.pdf

Becker, H. (1949). The nature and consequences of black propaganda. *American Sociological Review*, 14, 221–235. https://doi.org/10.2307/2086855.

Bennett, W. L., & Livingston, S. (2018). The disinformation order: Disruptive communication and decline of democratic institutions. *European Journal of Communication*, 33(2), 122–139. https://doi.org/10.1177/0267323118760317

Bertrand, N., & Herb, J. (2022, January 14). *US intelligence indicates Russia preparing operation to justify invasion of Ukraine*. CNN. https://edition.cnn.com/2022/01/14/politics/us-intelligence-russia-false-flag/index.html

Best, R. A., & Cumming, A. (Eds.). (2007). *Open Source Intelligence (OSINT): Issues for Congress*, CRS Report for Congress. Congressional Research Service. http://www.fas.org/sgp/crs/intel/RL34270.pdf

Bielska, A., Kurz, N.R., Baumgartner Y., & Benetis V. (Eds.). (2020). *Open Source Intelligence Tools and Resources Handbook 2020*. https://i-intelligence.eu/uploads/public-documents/OSINT_Handbook_2020.pdf

Boldyreva, E. L., Grishina, N. Y., & Duisembina, Y. (2018). Cambridge Analytica: Ethics and online manipulation with decision-making process. *The European Proceedings of Social & Behavioural Sciences*, 48, 91–102. https://doi.org/10.15405/epsbs.2018.12.02.10.

Boman, C. (2020). Examining characteristics of prebunking strategies to overcome PR disinformation attacks. *ScienceDirect*. https://doi.org/10.1016/j.pubrev.2021.102105

Borkowski, J. (2018). *Głupota. Kontrowersje i egzemplifikacje* [*Stupidity: controversies and exemplifications*]. Dom Wydawniczy Elipsa.

Bouchet, N. (2016). Russia's "militarization" of colour revolutions. *Policy Perspectives*, 4(2), 1–5.

Bowen, A. S. (2021). *Russian Military Intelligence: Background and Issues for Congress*. Congressional Research Service. https://crsreports.congress.gov/product/pdf/R/R46616/7

Bradshaw, S. (2020). *Influence operations and disinformation on social media*. Centre for International Governance Innovation. https://www.cigionline.org/articles/influence-operations-and-disinformation-social-media/

Brandt, J. (2021a). *How Democracies Can Win an Information Contest Without Undercutting Their Values*. Carnegie Endowment for International Peace. https://carnegieendowment.org/2021/08/02/how-democracies-can-win-information-contest-without-undercutting-their-values-pub-85058

Brandt, J. (2021b). How Autocrats Manipulate Online Information: Putin's and Xi's Playbooks. *The Washington Quarterly*. 44(3), 127–154. https://doi.org/10.1080/0163660X.2021.1970902

Brodnig, I. (2017). *Misinformation didn't change the outcome of the Bundestag election, but it still made headlines*. First Draft. https://www.firstdraftnews.org/latest/7-types-german-election/

Bryjka, F. (2014). Whistleblowing jako zagrożenie dla bezpieczeństwa informacyjnego państwa [Whistleblowing as a threat to state information security]. In T. Grabińska & H. Spustek (Eds.), *Bezpieczeństwo personalne a bezpieczeństwo strukturalne. Terroryzm i inne zagrożenia* [*Personal security versus structural security. Terrorism and other threats*] (pp. 95–119). WSOWL.

Bryjka, F. (2015). Cyberprzestrzeń w rosyjskiej strategii wojny hybrydowej [Cyberspace in Russian hybrid war strategy]. In T. Grabińska & Z. Kuźniar (Eds.), *Bezpieczeństwo personalne a bezpieczeństwo strukturalne. Czynniki antropologiczne i społeczne bezpieczeństwa personalnego* [*Personal security versus structural security. Anthropological and social factors of personal security*] (pp. 115–131). WSOWL.

Bryjka, F. (2016). *Rosyjska wojna zastępcza w Donbasie* [*Russian proxy war in Donbas*]. *Ante Portas – Studia nad Bezpieczeństwem*, 1, 201–219.

Bryjka, F. (2018). Rosyjskie środki aktywne w przestrzeni euroatlantyckiej [Russian active measures in the Euro-Atlantic space]. In T. Grabińska & Z. Kuźniar (Eds.), *Bezpieczeństwo personalne a bezpieczeństwo strukturalne* [*Personal security vs. structural security*] (pp. 168–180). AWL.

Bryjka, F. (2019). Rosyjskie środki aktywne w państwach Grupy Wyszehradzkiej [Russian active measures in Visegrad states]. In M. Banasik & A. Rogozińska (Eds.), *Różnorakie perspektywy bezpieczeństwa* [*Multiple perspectives of security*] (pp. 23–38). Difin.

Bryjka, F. (2021a). *Wojny zastępcze* [*Proxy wars*]. Polski Instytut Spraw Międzynarodowych.

Bryjka, F. (2021b). Białoruskie i rosyjskie działania dezinformacyjne wobec Polski w kontekście antyreżimowych protestów przeciwko Alaksandrowi Łukaszence [Belarusian and Russian disinformation activities against Poland in the context of anti-regime protests against Alyaksandr Lukashenko]. *Sprawy Międzynarodowe*, (4), 157–180. https://doi.org/10.35757/sm.2021.74.2.02

Bryjka, F. (2022). Russian Disinformation Regarding the Attack on Ukraine. *PISM Spotlight*, (15). https://www.pism.pl/publications/russian-disinformation-regarding-the-attack-on-ukraine.

Bryjka, F., & Legucka, A. (2021). Russian and Belarusian Disinformation and Propaganda in the Context of the Polish-Belarusian Border Crisis. *PISM Bulletin*, (212). https://www.pism.pl/publications/russian-and-belarusian-disinformation-and-propaganda-in-the-context-of-the-polish-belarusian-border-crisis

Bugayova, N., & Barros, G. (2020). *The Kremlin's Expanding Media Conglomerate*. Institute for the Study of War. https://www.understandingwar.org/backgrounder/kremlin%E2%80%99s-expanding-media-conglomerate

Bulao, J. (2020). *How Much Data Is Created Every Day in 2020*? TechJury. https://techjury.net/blog/how-much-data-is-created-every-day/

Buzan, B., Weaver, O., & de Wilde, J. (1998). *Security: A New Framework for Analysis*. Lynne Rienner.

Caddell, J. W. (2004). *Deception 101 – Primer on Deception*. U.S. Army War College.

Carmichael, F. (2021). *How a fake network pushes pro-China propaganda*. BBC News. https://www.bbc.com/news/world-asia-china-58062630

Carr, N. (2011). *The Shallows: What the Internet is Doing to Our Brains*. Norton & Company.

Carr, T. (2021). *Breaking Harmony Square. "No, nothing will be fine" – but could these misinformation games help at least a little?* Nieman Lab. https://www.niemanlab

.org/2021/08/no-nothing-will-be-fine-but-could-these-misinformation-games-help-at-least-a-little/

Carvin, S. (2021). *The Big Lie: Is Canada's Election 44 at Risk from Foreign Interference?* Centre for International Governance Innovation. https://www.cigionline.org/articles/the-big-lie-is-canadas-election-44-at-risk-from-foreign-interference/

Castells, M. (2011). *The Rise of the Network Society*. Wiley-Blackwell. https://doi.org/10.1002/9781444319514

Charen, M. (2003). *Useful Idiots: How Liberals Got It Wrong in the Cold War and Still Blame America First*. Regnery Publishing.

Charon, P., & Jeangene-Vilmer, J. B. (2021). *Chinese Influence Operations. A Machiavellian Moment*. Institute for Strategic Research.

Chiny wobec światowej pandemii COVID-19 [China facing a global pandemic COVID-19]. (2020). *Analizy OSW*. https://www.osw.waw.pl/pl/publikacje/analizy/2020-03-31/chiny-wobec-swiatowej-pandemii-covid-19

Chlebowicz, P. (2012). Interpretacja pojęcia dezinformacji w świetle art. 132 k.k. [Interpretation of the concept of disinformation in light of Article 132 of the Criminal Code]. *Studia Prawnoustrojowe*, (15), 41–48. https://wpia.uwm.edu.pl/czasopisma/sites/default/files/uploads/Studia_Prawno_Ustrojowe/2012/15/41-48.pdf

Chudy, W. (2003). *Filozofia kłamstwa* [*The philosophy of lying*]. Oficyna Wydawnicza Wolumen.

Clark, F. (2021). *Facebook confirms tests of a new anti-extremism warning prompt*. The Verge. https://www.theverge.com/2021/7/2/22560108/facebook-anti-extremism-prompt-user-resources-content-moderation

Clarke, R. M. (2010). *Intelligence Analysis: A target-centric approach*. CQ PR.

Clausewitz, C. von. (1984). *On War*. Princeton University Press.

Cole, R. (Ed.). (1998). *Encyclopedia of Propaganda*. Armonk: Sharpe Reference.

Collins, K. (2021). *Fake 5G coronavirus theories have real-world consequences*. CNET. https://www.cnet.com/tech/services-and-software/fake-5g-coronavirus-theories-have-real-world-consequences

Commin, G., & Filliol, E. (2015). Unrestricted warfare versus western traditional warfare: A comparative study. *Journal of Information Warfare*, 1, 14–23.

Conserva, H. T. (2003). *Propaganda Techniques*. 1st Books Library.

Constantinou, M., Kagialis, A., & Karekla, M. (2021). Is science failing to pass its message to people? Reasons and risks behind conspiracy theories and myths regarding COVID-19. *SSRN Electronic Journal*, (January), 1–10. https;//doi.org/10.2139/ssrn.3577662

Cooke, N. A. (2018). *Fake News and Alternative Facts: Information Literacy in a Post-Truth Era*. ALA Editions.

Corera, G. (2021). *Pro-Kremlin trolls target news website comments*. BBC News. https://www.bbc.co.uk/news/uk-58441662

Corlin, P., & Önnerfors, A. (2021). *Conspiracy theories are the staple diet of populism in Europe.* Voxeurop. https://voxeurop.eu/en/andreas-onnerfors-conspiracy-theories-are-the-staple-diet-of-populism-in-europe/

Cosentino, G. (2020). *Social Media and the Post-Truth World Order: The Global Dynamics of Disinformation.* Palgrave Macmillan.

Council of the European Union. (2018). Tackling online disinformation: A European approach. Brussels, 3 May 2018, https://data.consilium.europa.eu/doc/document/ST-8578-2018-INIT/en/pdf

Countering Information Influence Activities: A Handbook for Communicators (2019). Swedish Civil Contingency Agency. https://www.msb.se/RibData/Filer/pdf/28698.pdf

Courchesne, L., Ilhardt, J., & Shapiro, J. (2021, September). Review of social science research on the impact of countermeasures against influence operations. *Misinformation Review. Harvard Kennedy School.* https://misinforeview.hks.harvard.edu/article/review-of-social-science-research-on-the-impact-of-countermeasures-against-influence-operations/

CrossCheck France. (2021). First Draft. https://firstdraftnews.org/tackling/crosscheck/

Culliford, E. (2021a, July 13). *TikTok Sounds used to spread COVID vaccine misinformation – think tank.* Reuters. https://www.reuters.com/business/healthcare-pharmaceuticals/tiktok-sounds-used-spread-covid-vaccine-misinformation-think-tank-2021-07-13/

Culliford, E. (2021b, August 10). *Google restricts ad targeting of under-18s.* Reuters. https://www.reuters.com/technology/google-restricts-ad-targeting-under-18s-2021-08-10/

Culliford, E., & Potkin, F. (2021c, September 16). *Exclusive: Facebook cracks down on harmful real networks using playbook against fakes.* Reuters. https://www.reuters.com/technology/exclusive-facebook-target-harmful-coordination-by-real-accounts-using-playbook-2021-09-16/

Cunningham, S. B. (2002). *The Idea of Propaganda: A Reconstruction.* Praeger.

Da San Martino, G., Yu, S., Barrón-Cedeño, A., Petrov, R., & Nakov, P. (2019). Fine-Grained Analysis of Propaganda in News Article. In *Proceedings of the 2019 Conference on Empirical Methods in Natural Language Processing and the 9th International Joint Conference on Natural Language Processing* (pp. 5636–5646). Hong-Kong. https://www.aclanthology.org/D19-1.pdf

Dang, S., & Culliford, E. (2021). *Twitter sees jump in govt demands to remove content of reporters, news outlets.* Reuters. https://www.reuters.com/technology/exclusive-twitter-sees-jump-govt-demands-remove-content-journalists-news-outlets-2021-07-14/

Darcy, O. (2021). *Facebook takes action against 'disinformation dozen' after White House pressure.* CNN. https://edition.cnn.com/2021/08/18/tech/facebook-disinformation-dozen/index.html

Darczewska, J. (2014). *The anatomy of Russian information warfare. The Crimean operation, a case study*. Centre of Eastern Studies. www.osw.waw.pl/sites/default/files/the_anatomy_of_russian_information_warfare.pdf

Darczewska, J. (2015). *The devil is in the details. Information warfare in the light of Russia's military doctrine*. Centre of Eastern Studies. www.osw.waw.pl/en/publikacje/point-view/2015-05-19/devil-details-information-warfare-light-russias-military-doctrine

Darczewska, J. (2016). *Russia's armed forces on the information war front. Strategic documents*. Centre of Eastern Studies. www.osw.waw.pl/sites/default/files/prace_57_ang_russias_armed_forces_net.pdf

Darczewska, J. (2018). *Defenders of the besieged fortress On the historical legitimisation of Russia's special service*. Centre of Eastern Studies. https://www.osw.waw.pl/sites/default/files/pw_70_defenders-of-the-besieged-fortress_net.pdf

Darczewska, J., & Żochowski, P. (2017). Active measures. Russia's key export. Centre of Eastern Studies. www.osw.waw.pl/sites/default/files/pw_64_ang_active-measures_net_0.pdf

Davis, S. (2018). *Russian Meddling in Elections and Referenda in the Alliance*. NATO Parliamentary Assembly. https://www.nato-pa.int/download-file?filename=/sites/default/files/2018-11/181%20STC%2018%20E%20fin%20-%20RUSSIAN%20MEDDLING%20-%20DAVIS%20REPORT.pdf

Debunking Handbook. (2020). George Mason University. https://www.climatechangecommunication.org/debunking-handbook-2020/

Decker, T., & Boucher, T. (2021, July). *Disrupting online harms: A new approach*. The Global Disinformation Index. https://disinformationindex.org/wp-content/uploads/2021/07/2021-07-23-Disrupting-Online-Harms-A-New-Approach.pdf

DiResta, R. (2021, October 9). It's not misinformation. It's amplified propaganda. You don't need fake accounts to spread ampliganda online. Real people will happily do it. *The Atlantic*. https://www.theatlantic.com/ideas/archive/2021/10/disinformation-propaganda-amplification-ampliganda/620334/

DiResta, R., & Grossman, S. (2019). *Potemkin pages & personas: Assessing GRU online operations, 2014–2019*. Stanford University. https://cyber.fsi.stanford.edu/io/publication/potemkin-think-tanks

Disfavoured truths. Forgotten values. (2018). https://www.salwowski.net/2018/10/28/dwie-katolickie-definicje-klamstwa

Disinformation: EU assesses the Code of Practice and publishes platform reports on coronavirus related disinformation (2020). European Commission. https://www.ec.europa.eu/commission/presscorner/detail/en/ip_20_1568?utm_source=newsletter&utm_medium=email&utm_campaign=kremlin_watch_briefing_the_trump_team_welcomed_russian_disinformation&utm_term=2020-11-30

Disinformation risk assessment. The online news market in Brasil. Global Disinformation Index. (2021, September). https://disinformationindex.org/wp-content

/uploads/2021/09/GDI_Brazil-Disinformation-Risk-Assessment-Report-2021-ENGLISH.pdf

Disrupting Disinformation: A Global Snapshot of Government Initiatives. Global Disinformation Index. (2021, September). https://disinformationindex.org/wp-content/uploads/2021/09/2021-09-29-GDI-Global-Policy-Snapshot-Online.pdf

DiYanni, R. (2016). *Critical and Creative Thinking: A Brief Guide for Teachers*. Wiley-Blackwell.

Dizikes, P. (2021). *Study: Crowds can wise up to fake news. Experiment with Facebook-flagged content shows groups of laypeople reliably rate stories as effectively as fact-checkers do*. MIT. https://news.mit.edu/2021/crowd-source-fact-checking-0901

Domańska, M., & Rogoża, J. (2021). Forward, into the past! Russia's politics of memory in the service of 'eternal' authoritarianism. *OSW Report*. https://www.osw.waw.pl/sites/default/files/OSW-Report_Forward-into-the-past_net_0.pdf

Dubow, B. (2021, October 5). *Russia's rage reveals YouTube's strength*. CEPA. https://cepa.org/russias-rage-reveals-youtubes-strength/

Dubow, B., Lucas, E., & Morris, J. (2021, December 2). *Jabbed in the Back. Mapping Russian and Chinese Information Operations During COVID-19*. CEPA. https://cepa.org/comprehensive-reports/jabbed-in-the-back-mapping-russian-and-chinese-information-operations-during-the-covid-19-pandemic/

Duke Reporter's Lab. (2022). *Fact-checking*. https://reporterslab.org/fact-checking/

Dyner, A. M. (2020). On the Training Areas of the Union State of Belarus and Russia. *PISM Bulletin*, (203). https://www.pism.pl/publications/On_the_Training_Areas__of_the_Union_State_of_Belarus_and_Russia

Dyner, A. M., & Kacprzyk, A. (2022). U.S.-NATO Talks with Russia Yield No Breakthrough. PISM Bulletin, (3). https://www.pism.pl/publications/us-nato-talks-with-russia-yield-no-breakthrough

Edwards, A. (2021). *Agents of influence: Britain's secret intelligence war Against the IRA*. Merrion Press.

Ekman, P. (2009). *Telling Lies: Clues to Deceit in the Marketplace, Politics, and Marriage*. W. W. Norton & Company.

Elliott, V. (2021a, March 22). *Jordan's government used secretly recorded Clubhouse audio to spread disinformation*. Rest of World. https://restofworld.org/2021/jordan-clubhouse-tiktok-disinformation/

Elliott, V. (2021b, March 15). *"Disinformation influencers" for hire, only $15 a day. Rest of World*. https://restofworld.org/2021/kenya-disinformation-bbi-judiciary/

Eng, M. (2021, November 9). *What Illinois students will learn in media literacy class*. Axios. https://www.axios.com/local/chicago/2021/11/09/illinois-students-learn-media-literacy-class

Empowering fact-checkers *worldwide*. (2022). International Fact-Checking Network. https://www.poynter.org/ifcn/

Erbschloe, M. (2019). *Social Engineering Hacking Systems, Nations, and Societies.*CRC Press.

EU Code of Practice on Disinformation. (2018). https://www.ec.europa.eu/digital-single-market/en/news/code-practice-disinformation

EU should build a sanctions regime against disinformation. (2022, January). European Parliament. https://www.europarl.europa.eu/news/en/press-room/20220119IPR21313/eu-should-build-a-sanctions-regime-against-disinformation

European Democracy Action Plan: making EU democracies stronger. (2020, December). European Commission. https://www.ec.europa.eu/commission/presscorner/detail/en/ip_20_2250

European Democracy: Commission sets out new laws on political advertising, electoral rights and party funding. (2021, November 25). European Commission. https://ec.europa.eu/commission/presscorner/detail/en/ip_21_6118

European Federation of Journalists. (2021, August 31). *Malta: Journalists and public figures harassed in disinformation campaign.* https://europeanjournalists.org/blog/2021/08/31/malta-journalists-and-public-figures-harassed-in-disinformation-campaign/

Evaluating Information – Applying the CRAAP Test. (2017). Meriam Library, California State University. https://www.library.csuchico.edu/sites/default/files/craap-test.pdf

Exposing Russia's Effort to Sow Discord Online: The Internet Research Agency and Advertisements. (2018). US House of Representatives Permanent Selection Committee on Intelligence. https://intelligence.house.gov/social-media-content/

Exposing The Invisible. (2022). www.exposingtheinvisible.org/en/guides/google-dorking/

Facebook and academics row over data access. (2021, August 4). BBC News. https://www.bbc.com/news/technology-58086628

"Fake news" is Collins Dictionary's word of the year 2017. (2017). AP News. https://www.apnews.com/article/47466c5e260149b1a23641b9e319fda6

Ferraris, M. (2019). *Post verite et autores enigmes.* Puf.

Fitness app Strava lights up staff at military bases. (2018, January 29). BBC. https://www.bbc.com/news/technology-42853072

Foreign information manipulation and interference defence standards: Test for rapid adoption of the common language and framework 'DISARM'. (2022). Hybrid Centre of Excellence. https://www.hybridcoe.fi/publications/hybrid-coe-research-report-7-foreign-information-manipulation-and-interference-defence-standards-test-for-rapid-adoption-of-the-common-language-and-framework-disarm/

Foreign Service Information and Educational Exchange Circular 1951. Foreign Relations of the United States (1951, vol. 1). https://history.state.gov/historicaldocuments/frus1951v01/d333

Fox, M. (2021). *Vaccine Misinformation Is Spreading Among Kids, Too – Especially On Social Media*. Wisconsin Public Radio. https://www.wpr.org/node/1831796

Frankfurt, H.G. (2005). *On Bullshit*. Princeton University Press.

Fredheim, R. (Ed.). (2017–2021). *Robotrolling*. NATO Strategic Communications Centre of Excellence: Riga. https://stratcomcoe.org/publications?tid[]=8

Freeman, C.W. (2009). *Diplomat's Dictionary*. US Institute of Peace Press.

(2020, October 1). *French-Language Chinese State Media: Strategies and social media account analysis*. EU Disinfo Lab. https://www.disinfo.eu/publications/french-language-chinese-state-media-strategies-and-social-media-account-analysis/

Fukuyama, F. (2017). *The Emergence of a Post-Fact World*. Project Syndicate. https://www.project-syndicate.org/onpoint/the-emergence-of-a-post-fact-world-by-francis-fukuyama-2017-01

Fukuyama, F. (2021). Making the Internet safe for democracy, *Journal of Democracy*, 32(2), 37–44. https://doi.org/10.1353/jod.2021.0017

Fullinwieder, R. (2007). *Sissela Bok on lying and moral choice in private and public life – an amplification*. Infed. https://infed.org/mobi/sissela-bok-on-lying-and-moral-choice-in-private-and-public-life-an-amplification/

Fundacja Info Ops Polska. (2022). https://www.infoops.pl/

Galeotti, M. (2017, September 1). *Controlling Chaos: How Russia Manages Its Political War in Europe*. European Council on Foreign Relations. https://ecfr.eu/publication/controlling_chaos_how_russia_manages_its_political_war_in_europe/

Galeotti, M. (2019). *Russian Political War. Moving Beyond the Hybrid*. Routledge.

Galeotti, M. (2020). *The 'Gerasimov Doctrine' and Russian Non-Linear War, In Moscow Shadows*. Wordpress. https://www.inmoscowsshadows.wordpress.com/2014/07/06/the-gerasimov-doctrine-and-russian-non-linear-war/

Galloway, A. (2021 August, 9). Growing online 'influence-for-hire' economy opens door for foreign interference: Report. *The Sunday Morning Herald*. https://www.smh.com.au/politics/federal/growing-online-influence-for-hire-economy-opens-door-for-foreign-interference-report-20210809-p58h04.html

Garcia-Camargo, I., & Bradshaw, S. (2021). Disinformation 2.0: Trends for 2021 and beyond. *Hybrid Centre of Excellence Working Paper Series*. https://www.hybridcoe.fi/wp-content/uploads/2021/07/20210716_Hybrid_CoE_Working_Paper_11_Disinfo_2_0_WEB.pdf

GEC Special Report. (2020, August). *2020 Pillars of Russia's Disinformation and Propaganda Ecosystem*. U.S. Department of State. https://www.state.gov/wp-content/uploads/2020/08/Pillars-of-Russia%E2%80%99s-Disinformation-and-Propaganda-Ecosystem_08-04-20.pdf

Gerasimov, W. (2013, February 27). Cennost' nauki w priedwidienyi [The value of science is anticipation]. *Woyenno-promyshlennyi Journal*. https://www.vpk-news.ru/sites/default/files/pdf/VPK_08_476.pdf

Gerrits, A. W. M. (2018). Disinformation in international relations. How important is it? *Security and Human Rights*, (29), 3–23. https://doi.org/10.1163/18750230-02901007

Ghaffary, S., & Heilweil, R. (2021). *A new bill would hold Facebook responsible for Covid-19 vaccine misinformation*. Vox. https://www.vox.com/recode/2021/7/22/22588829/amy-klobuchar-health-misinformation-act-section-230-covid-19-facebook-twitter-youtube-social-media

Giles, K. (2016). Handbook of Russian information warfare. *NATO Defence College Fellowship Monograph*, 9. https://www.researchgate.net/publication/313423985_Handbook_of_Russian_Information_Warfare

Gilovich, T., Griffin, D., & Kahneman, D. (Eds.). (2012). *Heuristics and Biases: The Psychology of Intuitive Judgment*. Cambridge University Press.

Głowacka, D., Obem, A., & Szumańska, M. (2019). *Stop dezinformacji. Przewodnik dla dziennikarzy i redakcji* [*Stop disinformation. A guide for journalists and editors*]. Fundacja Panoptykon. https://panoptykon.org/stop-dezinformacji-przewodnik

Global Disinformation Index. (2021). *Ad-funded Anti-Semitism*. https://disinformationindex.org/wp-content/uploads/2021/07/DisinfoAds-English-Anti-Semitism-Disinformation.pdf

Gold, A. (2021). *Exclusive: New bipartisan bill takes aim at algorithms*. Axios. https://www.axios.com/algorithm-bill-house-bipartisan-5293581e-430f-4ea1-8477-bd9adb63519c.html

Golitsyn, A. (1984). *New Lies for Old: The Communist Strategy of Deception and Disinformation*. Dodd, Mead & Company.

Goliński, Z. (1936). Nauka Jana Kasjana o kłamstwie użytecznym [John Cassian's teaching on useful lying]. *Collectanea Theologica*, 17(4), 491–503.

Goodman, J., & Carmichael, F. (2020, July 26). *Coronavirus: False and misleading claims about vaccines debunked*. BBC. https://www.bbc.com/news/53525002.

Grabmeier, J. (2021). *Fact-checking works across the globe to correct misinformation. Study in four countries finds value in fighting false beliefs*. Ohio State News. https://news.osu.edu/fact-checking-works-across-the-globe-to-correct-misinformation/

Gribkov, A. I., & Smith, W. Y. (1993). *Operation Anadyr: US and Soviet Generals Recount the Cuban Missile Crisis*. Edition Q.

Gross, B. M. (1964). *The Managing of Organizations: The Administrative Struggle* (Vols. 1–2). Free Press.

Grotius, H. (1957). *O prawie wojny i pokoju. Trzy księgi, w których znajdują wyjaśnienie prawo natury i prawo narodów, a także główne zasady prawa publicznego* [*On the Law of War and Peace. Three books in which the law of nature and the law of nations are explained, as well as the main principles of public law*]. Państwowe Wydawnictwo Naukowe.

Guillois, A. (1857–1858). *Wykład historyczny, dogmatyczny, moralny, liturgiczny i kanoniczny wiary katolickiej z odpowiedziami na zarzuty wzięte z nauk przeciw religii, albo*

Teologia dogmatyczna i moralna, ku użyciu wiernych Chrystusowych (Vols. I–IV) [*A historical, dogmatic, moral, liturgical, and canonical lecture of the Catholic faith with answers to the objections taken from the sciences against religion, or Dogmatic and Moral Theology, for the use of the faithful of Christ*]. Druk. J. Glücksberga.

Gunitsky, S. (2021, April 21). Democracies can't blame Putin for their disinformation problem. *Foreign Policy*. https://foreignpolicy.com/2020/04/21/democracies-disinformation-russia-china-homegrown/

Hacker, P. M. S. (2018). *The Passions: A Study of Human Nature*. Wiley & Sons.

Haden, J. (2021). *Research Reveals How Many Likes It Takes for Facebook to Know You Better Than Your Spouse*. Inc. https://www.inc.com/jeff-haden/research-reveals-how-many-likes-it-takes-for-facebook-to-know-you-better-than-your-spouse.html

Hagey, K., & Horwitz, J. (2021, September 15). Facebook Tried to Make Its Platform a Healthier Place. *The Wall Street Journal*. https://www.wsj.com/articles/facebook-algorithm-change-zuckerberg-11631654215?st=vhgkxd1k8b70fle&reflink=desktop webshare_permalink

Hannah Arendt and the fragility of facts. (2021). EUvsDisinfo. https://euvsdisinfo.eu/hannah-arendt-and-the-fragility-of-facts/

Hanson, F., O'Connor, S., Walker, M., & Courtois, L. (2019, May). *Hacking democracies: Cataloguing cyber-enabled attacks on elections*. Australian Strategic Policy Institute. https://www.aspi.org.au/report/hacking-democracies

Hao, K. (2021a, September 16). *Troll farms reached 140 million Americans a month on Facebook before 2020 election, internal report shows*. MIT Technology Review. https://www.technologyreview.com/2021/09/16/1035851/facebook-troll-farms-report-us-2020-election/

Hao, K. (2021b, October 5). *The Facebook whistleblower says its algorithms are dangerous. Here's why*. MIT Technology Review. https://www.technologyreview.com/2021/10/05/1036519/facebook-whistleblower-frances-haugen-algorithms/

Harari, Y. (2018). *Homo Deus: A Brief History of Tomorrow*. Harper Collins Publ.

Harrel, Y. (2013). *La cyberstrategie russe*. Nuvis.

Hassan, N. A., & Hijazi, R. (2018). *Open Source Intelligence Methods and Tools: A Practical Guide to Online Intelligence*. APress.

Havlíček, P., & Yeliseyeu, A. (2021). *Disinformation Resilience Index in Central and Eastern Europe in 2021*. East Centre. https://east-center.org/wp-content/uploads/2021/09/DRI-report-2021.pdf

Hellerstein, E. (2021a, September 12). *The fevered world of antisemitic vaccine conspiracies*. Coda. https://www.codastory.com/disinformation/anti-semitism-anti-vaxxer/

Hellerstein, E. (2021b, September 24). *Information warfare is on the rise. Why aren't more people taking it seriously?* Coda. https://www.codastory.com/disinformation/information-warfare/

Helmus, T., & Keep, M. (2021). *A Compendium of Recommendations for Countering Russian and Other State-Sponsored Propaganda. RAND Corporation*. https://www.rand.org/pubs/research_reports/RRA894-1.html

Hendrix, J. (2021). *New report finds asymmetry in social media moderation favors dominant groups*. Tech Policy Press. https://techpolicy.press/new-report-finds-asymmetry-in-social-media-moderation-favors-dominant-groups/

Henry-Nickie, M., Frimpong, K., & Sun, H. (2019). *Trends in the information technology sector*. The Brookings Institution. https://www.brookings.edu/research/trends-in-the-information-technology-sector/#footnote-2

Herpen, M. van. (2015). *Putin's Propaganda Machine. Soft Power and Russian Foreign Policy*. Rowman & Littlefield Publishers.

Hobes, A. (2021). *In an Era of Misinformation and Tracking Technology, Long-Held Journalism Norms Are Shifting*. Nieman Reports. https://niemanreports.org/articles/in-an-era-of-misinformation-and-tracking-technology-long-held-journalism-norms-are-shifting/

Hoffman, F. (2007). *Conflict in the Twenty-First Century. The Rise of Hybrid Warfare*. Potomac Institute for Policy Studies.

Hoffman, F.G. (2009). Hybrid Warfare and Challenges. *Joint Force Quarterly*, 52(1), 34–39.

Hosaka, S. (2020). Repeating history: Soviet offensive counterintelligence active measures. *International Journal of Intelligence and CounterIntelligence*, 35(3). https://doi.org/10.1080/08850607.2020.1822100

Houston, P., Floyd, M., & Carnicero, S. (2012). *Spy the lie: Former CIA officers teach you how to detect deception*. St. Martin's Press.

How many people verified online information in 2021? (2021, December 16). EUROSTAT. https://ec.europa.eu/eurostat/web/products-eurostat-news/-/ddn-20211216-3

How To Spot Fake News. (2017, March). International Federation of Library Associations and Institutions. https://repository.ifla.org/handle/123456789/167

Howard, M. (1990). *British Intelligence in the Second World War* (Vol. 5: Strategic Deception). HMSO.

Howard, P. (2020). *Lie Machines*. Yale University Press.

Huang, B., & Carley, K. M. (2020). *Disinformation and Misinformation on Twitter During the Novel Coronavirus Outbreak*. Carnegie Mellon University. https://arxiv.org/pdf/2006.04278.pdf

Hubert, I., Shawn Eib, C., & Hirak, H. (2021). *Exposing a multi-year effort to manipulate Algeria's online political discourse and suppress dissident voices*. Graphika Report. https://graphika.com/reports/hammering-hirak/

Hughes, H. A., & Waismel-Manor, I. (2021). The Macedonian fake news industry and the 2016 US election. *Political Science & Politics*, January, 19–23. https://doi.org/10.1017/S1049096520000992

Hughes-Wilson, J. (2000). *Military Intelligence Blunders*. Da Capo Pres.

Huguet, A., Pane, J., Baker, G., Hamilton, L., & Faxon-Mills, S. (2021). *Media Literacy Education to Counter Truth Decay: An Implementation and Evaluation Framework*. RAND Corporation. https://www.rand.org/pubs/research_reports/RRA112-18.html

Hybrid CoE. (2022). *Hybrid Threats*. https://www.hybridcoe.fi/hybrid-threats/

Hybrid CoE. (2022). *Foreign information manipulation and interference defence standards: Test for rapid adoption of the common language and framework 'DISARM'* (Hybrid CoE Research Report 7). https://www.hybridcoe.fi/publications/hybrid-coe-research-report-7-foreign-information-manipulation-and-interference-defence-standards-test-for-rapid-adoption-of-the-common-language-and-framework-disarm/

International standards and comparative national approaches to countering disinformation in the context of freedom of the media. (2019). OSCE. https://www.osce.org/representative-on-freedom-of-media/424451

InVID. (2022). https://www.invid-project.eu/

Jackson, D. (2018). *How disinformation impacts politics and publics*. National Endowment for Democracy. https://www.ned.org/issue-brief-how-disinformation-impacts-politics-and-publics

Jakubowicz, M. (2016). Typy mediów [Media types]. In M. Graszewicz & M. Wszołek (Eds.), *Teorie komunikacji i mediów* [*Communication and media theories*] (pp. 61–101). Libron.

Jayakumar, S., Ang, B., & Anwar, N. D. (Eds.). (2021). *Disinformation and fake news*. Palgrave McMillan.

Jeangène, V. J. (2021a). Hybrid CoE Research Report 2: *Effective state practices against disinformation: Four country case studies*. https://www.hybridcoe.fi/publications/hybrid-coe-research-report-2-effective-state-practices-against-disinformation-four-country-case-studies/

Jeangène, V. J. (2021b). *Information defense: Policy measures taken against foreign information manipulation*. Atlantic Council. https://www.atlanticcouncil.org/wp-content/uploads/2021/07/Information-Defense-07.2021.pdf

Johnston, R. (2005). *Analytic culture in the U.S. intelligence community. An ethnographic study*. CIA Center for the Study of Intelligence.

Jones, P. (2021). *The conspiracy and disinformation challenge on e-commerce platforms*. Brookings. https://www.brookings.edu/techstream/the-conspiracy-and-disinformation-challenge-on-e-commerce-platforms/

Journalism, fake news, disinformation. Handbook for journalism education and training (2018). UNESCO.

Jowett, G., & O'Donnell, V. (2015). *Propaganda and persuasion*. Sage Publications.

Jurczyszyn, Ł. (2017). Russia's attempts to influence the presidential election in France. *PISM Bulletin*, (20). https://www.pism.pl/publications/Russia_s_Attempts_to_Influence_the_Presidential_Election_in_France (Accessed: 26 January 2022).

Juurvee, I., Sazonov, V., Parppei, K., Engizers, E., Pałasz, I., & Zawadzka, M. (2020). *Falsification of history as a tool of influence.* NATO Strategic Communications Centre of Excellence. https://stratcomcoe.org/publications/falsification-of-history-as-a-tool-of-influence/16

Kahneman, D. (2011). *Thinking fast and slow.* Penguin Books.

Kaplan, A. (2021). Gab's CEO is trying to use his platform to sabotage coronavirus vaccination efforts. Media Matters. https://www.mediamatters.org/coronavirus-covid-19/gabs-ceo-trying-use-his-platform-sabotage-coronavirus-vaccination-efforts

Karásková, I. (2020). *One China under media heaven: How Beijing hones its skills in information operations.* Hybrid Centre of Excellence Strategic Analysis, (23). https://www.hybridcoe.fi/wp-content/uploads/2020/06/20200625_Strategic-Analysis_23_China_Web.pdf

Kasapoglu, C. (2015). *Russia's renewed military thinking: Non-linear warfare and reflexive control.* NATO Defence College, Research Paper, (121). https://www.ndc.nato.int/news/news.php?icode=877

Kasapoglu, C., & Fery, M. (2020). *Iran's proxy war in Yemen: The information warfare landscape.* NATO Strategic Communications Centre of Excellence. https://www.stratcomcoe.org/irans-proxy-war-yemen-information-warfare-landscape

Kavanagh, J., & Rich, M.D. (2018). *Truth decay: An initial exploration of the diminishing role of facts and analysis in American public life.* RAND Corporation. https://www.rand.org/pubs/research_reports/RR2314.html

Kaye, B. (2021). *CNN denies Australians access to its Facebook pages, cites defamation risk.* Reuters. https://www.reuters.com/technology/cnn-quits

Kayyali, D., & York, J. (2021). *The Facebook Oversight Board is making good decisions – but does it matter?* Tech Policy Press. https://techpolicy.press/the-facebook-oversight-board-is-making-good-decisions-but-does-it-matter/

Keane, J. (1991). *Media and democracy.* Polity.

Kennan, G. (1948). *Policy Planning Staff Memorandum.* US Department of State, Office of the Historian. https://archive.law.upenn.edu/live/files/9964-kennan-memo-political-warfarepdf

Kessler, G., Rizzo, S., & Kelly, M. (2021, January 24). Trump's false or misleading claims total 30,573 over 4 years. *The Washington Post.* https://www.washingtonpost.com/politics/2021/01/24/trumps-false-or-misleading-claims-total-30573-over-four-years/

Kick, R. (Ed.). (2001). *The disinformation guide to media distortion, historical whitewashes and cultural myths.* Disinformation Network.

Killeen, M. (2021a, October 13). *EU Commission gathers expert group on disinformation and digital literacy.* Euractive. https://www.euractiv.com/section/digital/news/eu-commission-gathers-expert-group-on-disinformation-and-digital-literacy/

Killeen, M. (2021b, October 18). *New signatories to join EU anti-disinformation code amid calls for improvement.* Euractive. https://www.euractiv.com/section/digital/news/new-signatories-to-join-eu-anti-disinformation-code-amid-calls-for-improvement/

Kimball, W. (2021, August 25). *Facebook news consumers are more anti-vaccine than Fox News viewers, study finds.* Gizmodo. https://gizmodo.com/facebook-news-consumers-are-more-anti-vaccine-than-fox-1847378398

Kirkham, R. L. (1992). *Theories of Truth: A Critical Introduction.* MIT Press.

Kissinger, H. (1969). *Nuclear Weapons and Foreign Policy.* W.W. Norton & Co.

Knight, T. (2021). *Tools + Tech: How DFRLab cracks cases of disinformation. Medium Code for Africa.* https://medium.com/code-for-africa/tools-tech-how-dfrlab-cracks-cases-of-disinformation-5ab97be33a96

Kobla, M. (2019). *Notre Dame: jak zmanipulowano zdjęcie dwójki uśmiechniętych mężczyzn? Kłamstwa wokół pożaru* [*Notre Dame: how was a photo of two smiling men manipulated? Lies surrounding the fire*]. AntyFake. https://www.antyfake.pl/notre-dame-muzulmanie-fake-news

Kochan, M. (2007). *Pojedynek na słowa: Techniki erystyczne w publicznych sporach* [*Duel of words: eristic techniques in public disputes*]. Wydawnictwo Znak.

Kolesnyk, D. (2021). *France unveils information operations doctrine.* Military Technology. https://kolesnyk.fr/images/miltech62021_franceL2I_kolesnyk.pdf

Kołodziej, J. H. (2017). Narratologia w badaniach komunikacji politycznej: Metodologiczne przymiarki [Narratology in political communication research: Methodological adaptations]. *Polityka i Społeczeństwo*, (1), 26. https://doi.org/10.15584/polispol.2017.1.2

Konstytucja Rzeczypospolitej Polskiej z dnia 2 kwietnia 1997 r. [*Constitution of the Republic of Poland of April 2, 1997*] (1997). http://isap.sejm.gov.pl/isap.nsf/download.xsp/WDU19970780483/U/D19970483Lj.pdf

Korolko, M. (2010). *Podręcznik retoryki homiletycznej* [*Handbook of homiletical rhetoric*]. Wydawnictwo WAM.

Kowalska, M., & Wigienka, S. (2020). *StratCom: perspektywa polska. Struktura systemu komunikacji i analiza kampanii na temat 20. rocznicy przystąpienia Polski do NATO* [*StratCom: a Polish perspective. Structure of the communication system and analysis of the campaign on the 20th anniversary of Poland's accession to NATO*]. Centrum Analiz Propagandy i Dezinformacji. https://capd.pl/pl/analizy/221-stratcom-perspektywa-polska-struktura-systemu-komunikacji-i-analiza-kampanii-na-temat-20-rocznicy-przystapienia-polski-do-nato

Kozłowska, A. (2016). Wpływ mass mediów na życie społeczne [Impact of mass media on social life]. In E. Firlit & J. Gładys-Jakóbik (Eds.), *Wybrane problemy współczesnego świata w refleksji socjologicznej* [*Selected problems of the modern world in sociological reflection*] (pp. 195–215). Szkoła Główna Handlowa.

Kruger, J., & Dunning, D. (1999). Unskilled and unaware of it: How difficulties in recognizing one's own incompetence lead to inflated self-assessments. *Journal of Personality and Social Psychology*, 77(6), 1121–1134. https://doi.org/10.1037/0022-3514.77.6.1121.

Kruglashov, A., & Shvydiuk, S. (2020). Hybrydowe zagrożenia dla demokracji. Wybrane przykłady zewnętrznej ingerencji Rosji w wybory [Hybrid threats to democracy. Selected examples of Russia's external interference in elections]. *Wschód Europy*, 2, 79–93. https://doi.org/10.17951/we.2020.6.2.79-93.

Kucharski, J. (2014). *Usprawiedliwione kłamstwo we współczesnej etyce stosowanej [Justifiable lying in modern applied ethics]*. Wydawnictwo WAM & Akademia Ignatianum.

Kuenne, W. (2003). *Conception of Truth*. Oxford: Clarendon Press.

Kuo, R., & Marwick, A. (2021). *Critical disinformation studies: History, power, and politics*. Harvard Kennedy School. https://misinforeview.hks.harvard.edu/article/critical-disinformation-studies-history-power-and-politics/#.YRTb1vew_e8.twitter

Kupiecki, R. (1993). *Natchnienie milionów. Kult Józefa Stalina w Polsce 1944–1956 [The inspiration of millions. The cult of Joseph Stalin in Poland 1944–1956]*. Wydawnictwa Szkolne i Pedagogiczne.

Kupiecki, R. (2015). The Meaning of Military Victory. In Search of a New Analytical Framework. *Security and Defence Quarterly*, 2(3), 7–28. https://doi.org/10.5604/23008741.1152567

Kupiecki, R. (2019). Mit założycielski polityki zagranicznej Rosji [The founding myth of Russian foreign policy]. *Sprawy Międzynarodowe*, 72(4), 77–105. https://doi.org/10.35757/SM.2019.72.4.03

Kupiecki, R. (2020a). Sztuczna inteligencja a bezpieczeństwo międzynarodowe w przyszłości [Artificial intelligence and future international security]. In R. Kuźniar, A. Bieńczyk-Missala, P. Grzebyk, R. Kupiecki, M. Madej, K. Pronińska, A. Szeptycki, P. Śledź, M. Tabor, & A. Wojciuk (Eds.), *Bezpieczeństwo międzynarodowe [International security]* (pp. 472–497). Wydawnictwo Naukowe Scholar.

Kupiecki, R., & Menkiszak, M. (Eds.) (2020b). *Documents Talk. NATO–Russia Relations after the Cold War*. Polish Institute of International Affairs.

Kupiecki, R., Chłoń, T., Bryjka, F., Kozłowski, K., Misiuna, J., Podemska, J., & Podemski, P. (2021). *Platforma zwalczania dezinformacji. Budowanie odporności społecznej: badania i edukacja [A platform for combating disinformation. Building public resilience through research and education]*. Dom Wydawniczy Elipsa.

Kupiecki R. (2022a). Western betrayal. The founding myth of Russian foreign policy. In A. Legucka, & R. Kupiecki. *Disinformation, Narratives and Memory Politics in Russia and Belarus* (pp. 43–58). Routledge. https://doi.org/10.4324/9781003281597-4

Kupiecki R. (2022b). Poland's security policy. *Siyasal: Journal of Political Sciences*, 31(Special Issue). http://dx.doi.org/10.26650/siyasal.2022.31.945221

Kuzichkin, A., & Hanley, M. (2021). *Russian media landscape. Structures, mechanisms, and technologies of information Operations*. NATO Strategic Communications Centre of Excellence. https://stratcomcoe.org/publications/russian-media-landscape-structures-mechanisms-and-technologies-of-information-operations/215

LaFrance, A. (2021). *The largest autocracy on earth. Facebook is acting like a hostile foreign power; it's time we treated it that way*. The Atlantic. https://www.theatlantic.com/magazine/archive/2021/11/facebook-authoritarian-hostile-foreign-power/620168/

Lanoszka, A. (2019). Disinformation in international politics. *European Journal of International Security*, 4(2), 227–248. https://doi.org/10.1017/eis.209.6

Lasswell H.D. Propaganda Technique in the World War. (2013). Martino Fine Books.

Lasswell H.D., & Blumenstock D. World Revolutionary Propaganda: A Chicago Study. (1939). New York, Knopf.

Lasswell H.D., & Smith B.C., & Casey R.D. Propaganda, Communication, and Public Opinion: A Comprehensive Reference Guide. (1946). Princeton University Press.

Lasswell H.D., & Lerner D. Speier H. (Eds.). Propaganda and Communication in World History. (1980). Vol. 1–3. The University Press of Hawaii.

Legucka, A. (2019). Countering Russian Disinformation in the European Union. *PISM Bulletin* 111. https://pism.pl/publications/Countering_Russian_Disinformation_in_the_European_Union

Legucka, A. (2020). How Not to be a Useful Idiot in Relations with Russia. *New Eastern Europe*, 1–2. https://www.neweasterneurope.eu/2020/01/28/how-not-to-be-a-useful-idiot-in-relations-with-russia

Legucka, A. (2021a). Russia Demands Security Guarantees from the U.S. and NATO. *PISM Bulletin* 214. https://www.pism.pl/publications/russia-demands-security-guarantees-from-the-us-and-nato

Legucka, A. (2021b). Online Warfare: Russian Policy on International Information Security. *PISM Bulletin* 111. https://www.pism.pl/publications/Online_Warfare_Russian_Policy_on_International_Information_Security

Legucka, A. (2022). Russian Society on Their Country's Invasion of Ukraine. *PISM Spotlight 18*. https://www.pism.pl/publications/russian-society-on-their-countrys-invasion-of-ukraine

Legucka, A., & Kupiecki, R. (Eds). (2022). Disinformation, Narratives and Memory Politics in Russia and Belarus. Routledge. https://doi.org/10.4324/9781003281597

Legucka, A., & Przychodniak, M. (2020). Disinformation from China and Russia during the COVID-19 pandemic. *PISM Bulletin 86*. https://www.pism.pl/publications/Disinformation_from_China_and_Russia__during_the_COVID19_Pandemic

Lennon, A. (Ed.). (2003). *The Battle for Hearts and Minds Using Soft Power to Undermine Terrorist Networks*. Washington Quarterly Readers.

Levin, D. H. (2020). *Meddling in the Ballot Box: The Causes and Effects of Partisan Electoral Interventions*. Oxford University Press.

Levine, T. R. (Ed.). (2014). *Encyclopedia of Deception*. Sage.

Levitin, D. (2016). *Weaponized Lies: How to Think Critically in the Post-Truth Era*. Penguin Audio.

Lewis, M. (2016). *The Undoing Project: A Friendship That Changed Our Minds*. W. W. Norton & Company.

Liang, Q., & Xiangsui, W. (2002). *Unrestricted Warfare: China's Master Plan to Destroy America*. Panamerican Publishing Company.

Liedel, K., Piasecka, P., & Aleksandrowicz, T. R. (2012). *Analiza informacji: Teoria i praktyka* [*Information analysis in theory and practice*]. Difin.

Lorenz, W. (2021). *Odstraszanie. Strategia i polityka* [Deterrence; Strategy and politics]. Polski Instytut Spraw Międzynarodowych.

Loucaides, D. (2021, September 5). *Who is behind Spanish Telegram's storm of Covid-19 disinformation? A complex web of fake news and foreign propaganda has fueled vaccine skepticism and anti-lockdown riots*. Coda. https://www.codastory.com/disinformation/spain-telegram-covid19-disinformation/

Machiavelli, N. (2014). *The Prince*. Penguin Books.

Mann, I. (2008). *Hacking the human: Social engineering techniques and security countermeasures*. Gover.

Martin, J. (1958). *International Propaganda. Its Legal and Diplomatic Control*. University of Minnesota Press.

Materska, K. (2007). *Informacja w organizacjach społeczeństwa wiedzy* [*Information in knowledge society organizations*]. Wydawnictwo Stowarzyszenia Bibliotekarzy Polskich.

Mattis, P. (2018). *China's three warfares' in perspective. War on the Rocks*. https://www.warontherocks.com/2018/01/chinas-three-warfares-perspective

Maurer, T. (2018). Cyber proxies and their implications for liberal democracies. *The Washington Quarterly*, 41(2), 171–188. https://doi.org/10.1080/0163660X.2018.1485332.

Mazarr, M. J., Casey, A., Demus, A., Matthews, L.J., Beauchamp-Mustafaga, N.,, Sladden, J. (2019). *Hostile Social Manipulation. Present Realities and Emerging Trends*. RAND Corporation. https://www.rand.org/pubs/research_reports/RR2713.html

McBride, M., Wolters, W., Haney, K., & Rosenau, W. (2021). *The Psychology of (Dis)information: Case Studies and Implications*. CAN.

McDougall, J., Zezulkova, M., van Driel, B., & Sternadel, D. (2018). *Teaching media literacy in Europe: Evidence of effective school practices in primary and secondary education*. Publications Office of the European Union. https://doi.org/10.2766/613204. https://www.researchgate.net/publication/329718142_Teaching_media_literacy_in_Europe_evidence_of_effective_school_practices_in_primary_and_secondary_education

McInnis, K., & Starling, C. (2021, June). *The Case for a Comprehensive Approach 2.0: How NATO Can Combat Chinese and Russian Political Warfare*. Atlantic Council. https://

www.atlanticcouncil.org/wp-content/uploads/2021/06/NATO-Comprehensive-Approach-Report-2021.pdf

McIntyre, L. (2018). *Post-Truth*. MIT Press.

McLuhan, M. (1970). *Culture is our business*. Wipf and Stock Publishers.

McLuhan, M. (1994). *Understanding media: The extensions of man*. MIT Press.

McNair, B. (2017). *Fake News: Falsehood, Fabrication and Fantasy in Journalism (Disruptions)*. Routledge.

Media Market Risk Ratings: Malaysia. (2021, July). Global Disinformation Index. https://disinformationindex.org/wp-content/uploads/2021/07/2021-06-15-Malaysia-Risk-Ratings-Report-Online.pdf

Melford, C. & Rogers, D. (2021, July 29). *Want Less Awful Content? Stop Focusing on Content Moderation*. Global Disinformation Index. https://disinformationindex.org/2021/07/want-less-awful-content-stop-focusing-on-content-moderation/

Melton, H. K., & Wallace, R. (2010). *The Official CIA Manual of Trickery and Deception*. Harper.

Menkiszak, M. (2016, November 9). Moscow chooses Trump. Russia on the US presidential elections. *OSW Commentary* 277. Retrieved from https://www.osw.waw.pl/en/publikacje/osw-commentary/2016-11-09/moscow-chooses-trump-russia-us-presidential-elections

Menkiszak, M. (2021, December 20). *Russia's blackmail of the West*. Center for Eastern Studies. https://www.osw.waw.pl/en/publikacje/analyses/2021-12-20/russias-blackmail-west

Metas (formerly Facebook Inc.) annual revenue from 2009 to 2021. (2022). Statista. https://www.statista.com/statistics/268604/annual-revenue-of-facebook/

Metz, R. (2008, March 3). Fake news spreads faster than the truth, and it's all our fault. *MIT Technology Review*, March. https://www.technologyreview.com/2018/03/08/144839/fake-news-spreads-faster-than-the-truth-and-its-all-our-fault/

Migalski, M. & Kaczmarzyk, M. (2020). *Homo politicus sapiens. Biologiczne aspekty politycznej gry [Biological aspects of the political game]*. Wydawnictwo Sonia Draga.

Mitrochin, W. (2002). KGB Lexicon: The Soviet intelligence officer's handbook. London: Routledge.

Modus trollerandi (Parts 1–7). (2021). EUvsDisinfo. https://www.euvsdisinfo.eu/?s=trollerandi.

Najzer, B. (2020). *The hybrid age. International security in the era of hybrid warfare*. Bloomsbury Publishing.

Nardi, B. & O'Day, V. L. (1999). *Information ecologies: Using technology with heart*. MIT Press.

National threat assessment 2021. (2021). https://www.vsd.lt/wp-content/uploads/2021/03/2021-EN-el_.pdf

NATO *Military Policy on Strategic Communications*. (2017). https://www.stratcom.nuou.org.ua/wp-content/uploads/2020/01/NATO-MILITARY-POLICY-ON-STRATEGIC-COMMUNICATIONS.pdf

NATO. (2018). MC 0422/6 NATO Military Policy for Information Operations. NATO Unclassified, 11.08.2018.

NATO Open Source Intelligence Handbook. (2021, November). NATO SACLANT Intelligence Branch. https://www.academia.edu/4037348/NATO_Open_Source_Intelligence_Handbook

NATO 2030: United for a New Era. (2020, November). https://www.nato.int/nato_static_fl2014/assets/pdf/2020/12/pdf/201201-Reflection-Group-Final-Report-Uni.pdf

NATO's approach to countering disinformation. (2020). https://www.nato.int/cps/en/natohq/177273.htm

NATO. (2022). *Madrid Summit Declaration*. https://www.nato.int/cps/en/natohq/official_texts_196951.htm?selectedLocale=en

NATO 2022 Strategic Concept. (2022). https://www.nato.int/nato_static_fl2014/assets/pdf/2022/6/pdf/290622-strategic-concept.pdf

NATO-Russia: Setting the Record Straight. (2022a). https://www.nato.int/cps/en/natohq/115204.htm#myths.

NATO. (2022b). NATO 2022 Strategic Concept. Retrieved July 10, 2022, from https://www.nato.int/nato_static_fl2014/assets/pdf/2022/6/pdf/290622-strategic-concept.pdf

NATO. (2022c). Madrid Summit Declaration. Retrieved July 10, 2022, from https://www.nato.int/cps/en/natohq/official_texts_196951.htm?selectedLocale=en

NATO. (2022d). *Statement by NATO heads of state and government*. https://www.nato.int/cps/en/natohq/official_texts_193719.htm?selectedLocale=en

Nemr, Ch., & Gangwar, W. (2019, March). *Weapons of mass distraction. Foreign state-sponsored disinformation in the digital age*. Park Advisors. https://www.state.gov/wp-content/uploads/2019/05/Weapons-of-Mass-Distraction-Foreign-State-Sponsored-Disinformation-in-the-Digital-Age.pdf

Newman, N., Fletcher, R., Schulz, A., Andi, S., & Nielsen, R.K. (2020). *Reuters Institute Digital News Report 2020*. Reuters Institute for the Study of Journalism. https://www.reutersinstitute.politics.ox.ac.uk/sites/default/files/2020-06/DNR_2020_FINAL.pdf

Nichols, T. (2017). *The death of expertise: The campaign against established knowledge and why it matters*. Oxford University Press.

Nimmo, B. (2015). *Backdating the blame: How Russia made NATO a party to the Ukraine conflict*. NATO Strategic Communications Centre of Excellence. https://stratcomcoe.org/publications/backdating-the-blame-how-russia-made-nato-a-party-to-the-ukraine-conflict/178

Ociepka, B. (2002). *Komunikowanie międzynarodowe* [*International communication*]. Wydawnictwo Astrum.

Ogrodowczyk, A., Borkowska, M., Murawska-Najmiec, E., & Twardowska, K. (2020). *Fake news, dezinformacja online: próby przeciwdziałania tym zjawiskom z perspektywy instytucji międzynarodowych oraz wybranych państw UE, w tym Polski* [*Fake news, online disinformation: attempts to counter these phenomena from the perspective of international institutions and selected EU countries, including Poland*]. Departament Strategii Biura KRRiT. https://www.gov.pl/web/krrit/fake-news--dezinformacja-online

Oniszczuk, Z., & Gierula, M. (Eds.). (2007). *Mało znane systemy medialne* [*Little-known media systems*]. Wydawnictwo Humanitas.

Orr, C. (2021, July 29). Digital astroturfing campaign amplifies Fox News-inspired GB News. *Byline Times*. https://bylinetimes.com/2021/07/29/automated-culture-wars-digital-astroturfing-campaign-amplifies-fox-news-inspired-gb-news/

Orsek, B., & Ozsoy, F. F. (2020). *IFCN's Code of Principles Transparency Report for 2020*. Poynter. https://ifcncodeofprinciples.poynter.org/know-more/code-of-principles-1st-year-a-report

Osmann, J., Selva, M., & Feinstein, A. (2021). How have journalists been affected psychologically by their coverage of the COVID-19 pandemic? *BMJ Open*, 11(7), e045675. https://doi.org/10.1136/bmjopen-2020-0

Ostaszewski, A. (2018). Rodzaje kłamstwa i ich wykorzystanie w tekście [Types of lying and their use in a text]. *Rocznik Prasoznawczy*, 12, 33–53.

Pacepa, I. M., & Rychlak, R. J. (2013). *Disinformation: Former spy chief reveals secret strategies for undermining freedom, attacking religion, and promoting terrorism*. WND Books.

Pariser, E. (2011). *The Filter Bubble: What The Internet Is Hiding From You*. Penguin Press.

Paul, Ch., & Matthews, M. (2017). *The Russian 'Firehouse of Falsehood' propaganda model. Why it might work and options to counter it*. RAND Corporation. https://www.rand.org/pubs/perspectives/PE198.html/

Paul, K. (2021, September 21). Internet freedom on the decline in US and globally, study finds. *The Guardian*. https://www.theguardian.com/technology/2021/sep/21/internet-freedom-decline-free-speech-study

Paul, R., & Elder, L. (2009). *Miniature guide to critical thinking: Concepts and tools*. Foundation for Critical Thinking.

Paul, R., & Elder, L. (2014). *Critical thinking: Tools for taking charge of your learning and your life*. Pearson Education Limited.

Petraitis, D., Ratsiborynska, V., & Kirdemir, B. (2021). *Exercise Kavkaz 2020 – A Final Test of Russian Military Reform*? NATO Strategic Communications Centre of Excellence.

https://stratcomcoe.org/publications/exercise-kavkaz-2020-a-final-test-of-russian-military-reform/4

Philips, T. (2019). *Truth. A Brief History of Bullshit.* Wildfire.

Piotrowski, M. A. (2015). Konflikt nigdy nie jest prosty: amerykańska teoria i doktryna wojen oraz przeciwników hybrydowych [Conflict Is Never Simple: American Theory and Doctrine of Wars and Hybrid Opponents]. *Sprawy Międzynarodowe*, 69(2), 7–38.

Piotrowski, M. A. (2017). Intelligence Reports on Russian Interference in the U.S. Presidential Election. *PISM Bulletin* 8, https://pism.pl/publications/Intelligence_Reports_on_Russian_Interference_in_the_U_S__Presidential_Election

Pipe, E. (2021). *New Telegram Research Shows QAnon 'Largest Extremist Group' Online.* Logically. https://www.logically.ai/articles/new-telegram-research-shows-qanon-largest-extremist-group-online

Pismo Święte i tradycja kościoła o kłamstwie. Zbiór wypowiedzi [*The Holy Scripture and Tradition of the Church on Lies: A Collection of Statements*]. (2017). https://www.salwowski.net/2017/07/03/pismo-swiete-i-tradycja-kosciola-o-klamstwie-zbior-wypowiedzi/ (Accessed: 9.01.2022).

Platforma edukacyjna Demagog [*Demagog Educational Platform*]. (2022). Stowarzyszenie Demagog [Demagog Association], https://platforma.demagog.org.pl/ (Accessed: 31.03.2022).

Pollet, M. (2021). *Facebook bans Taliban but Twitter adopts more 'laissez faire' approach.* Euractiv, https://www.euractiv.com/section/digital/news/facebook-bans-taliban-but-twitter-adopts-more-laissez-faire-approach/

Pomerantsev, P. (2019). *This is not propaganda.* New York: Public Affairs.

Pomerantsev, P., & Weiss, M. (2014). *The menace of unreality. How the Kremlin weaponizes information, culture and money.* New Institute of Modern Russia, https://www.imrussia.org/media/pdf/Research/Michael_Weiss_and_Peter_Pomerantsev__The_Menace_of_Unreality.pdf

Posard, M., Reininger, H., & Helmus, T. (2021). *Countering foreign interference in U.S. election.* RAND Corporation. https://www.rand.org/content/dam/rand/pubs/research_reports/RRA700/RRA704-4/RAND_RRA704-4.pdf

Projekt doktryny bezpieczeństwa informacyjnego RP z 2015 r. [Draft informaction security doctrine 2015] (2015). Biuro Bezpieczeństwa Narodowego. https://www.bbn.gov.pl/ftp/dok/01/Projekt_Doktryny_Bezpieczenstwa_Informacyjnego_RP.pdf

Public consultation: Draft Principles of Good Practice for Public Communication Responses to Mis- and Disinformation (2022). OECD. https://www.oecd.org/gov/open-government/public-consultation-draft-principles-good-practice-public-communication-responses-to-mis-and-disinformation.html

Putin Chef's Kisses of Death: Russia's Shadow Army's State-Run Structure Exposed (2020). Bellingcat. https://www.bellingcat.com/news/uk-and-europe/2020/08/14/pmc-structure-exposed/

Putin's declaration of war on Ukraine (2022, February 24). *The Spectator*. https://www.spectator.co.uk/article/full-text-putin-s-declaration-of-war-on-ukraine

Questions and answers – The EU steps up action against disinformation (2018). European Commission. https://www.ec.europa.eu/commission/presscorner/detail/en/MEMO_18_6648

Radin, A., Demus, A., & Marcinek, K. (2020). *Understanding Russian Subversion: Patterns, Threats, and Responses*. RAND Corporation. https://www.rand.org/content/dam/rand/pubs/perspectives/PE300/PE331/RAND_PE331.pdf

Rajczyk, R. (2016). *Nowoczesne wojny informacyjne* [*Modern infowars*]. Warszawa: Difin.

Rapid Alert System. Strengthening Coordination and Joint Responses to Disinformation (2019). European External Action Service. https://www.eeas.europa.eu/sites/eeas/files/ras_factsheet_march_2019_0.pdf

Rashkin, H., Choi, E., Jang, Y.J., Volkova, S., & Choi, Y. (2017). Truth of Varying Shades: Analyzing Language in Fake News and Political Fact-Checking. In *Proceedings of the 2017 Conference on Empirical Methods in Natural Language Processing* (pp. 2931–2937). https://www.aclanthology.org/D17-1317.pdf

Ratsiborynska, V., Petraitis, D., & Akimenko, V. (2020). *Russia's Strategic Exercises: Messages and Implications*. NATO Strategic Communications Centre of Excellence. https://stratcomcoe.org/publications/russias-strategic-exercises-messages-and-implications/30

Report On The Investigation Into Russian Interference in The 2016 Presidential Election. (2019). US State Department. https://www.justice.gov/archives/sco/file/1373816/download

Ressa, M. (2016). *How Facebook algorithms impact democracy*. Rappler. https://www.rappler.com/newsbreak/facebook-algorithms-impact-democracy

Reuters Fact Check. (2021). *Fact Check-Photo shows 2018 protests in France, not 2021*. https://www.reuters.com/article/factcheck-france-protest-idUSL1N2OS1V2

Ricks, T.E. (2015, December 3). Narratives are about 'meaning', not 'truth'. *Foreign Policy*. https://www.foreignpolicy.com/2015/12/03/narratives-are-about-meaning-not-truth

Rid, T. (2020). *Active Measures: The Secret History of Disinformation and Political Warfare*. Farrar Strauss & Giroux.

Roberts, D. (2020). *China's Disinformation Strategy. Its Dimensions and Future*. Atlantic Council. https://www.atlanticcouncil.org/wp-content/uploads/2020/12/CHINA-ASI-Report-FINAL-1.pdf

Robinson, L., Helmus, T.C., Cohen, R.S., Nader, A., Radin, A., Magnusson, M., & Migacheva, K. (2018). *Modern Political Warfare. Current Practices and Possible Responses*. RAND Corporation.

Rodgers, J., & Lanoszka, A. (2021). Russia's rising military and communication power: From Chechnya to Crimea. *Media, War & Conflict*, August, 1–18. https://doi.org/10.1177/17506352211027084

Roncone, G., Wahlstrom, A., Revelli, A., Mainor, D., Riddell, S., & Read, B. (2021). *UNC1151 Assessed with High Confidence to have Links to Belarus, Ghostwriter Campaign Aligned with Belarusian Government Interests*. Mandiant. https://www.mandiant.com/resources/unc1151-linked-to-belarus-government

Rosińska, K. (2021). *Fake news. Geneza, istota, przeciwdziałanie* [*Fake news. Its origins, essence and counteraction*]. Wydawnictwo Naukowe PWN.

Rusbridger, J. (1989). *Intelligence Game*. Random House UK.

Russia Active Measures Campaign and Interference in the 2016 US Election. Report. (2020). Vol. 1–5. US Senate Intelligence Committee. https://www.intelligence.senate.gov/publications/report-select-committee-intelligence-united-states-senate-russian-active-measures

Russia's top five myths about NATO. (2023). https://www.nato.int/nato_static_fl2014/assets/pdf/2021/10/pdf/2110-russia-top5-myths-en.pdf.

Sample, C., Jensen, M.J., Scott, K., McAlaney, J., Fitchpatrick, S., Brockinton, A., & Ormrod, D. (2020, December 16). Interdisciplinary lessons learned while researching fake news. *Frontiers in Psychology*. https://doi.org/10.3389/fpsyg.2020.537612

Sands, G. (2021). *Homeland Security warns of potential conspiracy theory-fueled violence in August*. CNN. https://edition.cnn.com/2021/08/06/politics/homeland-security-warning-political-conspiracy-threat-violence/index.html

Schaer, C. (2021). *How Hamburg became Europe's unlikely data protection trailblazer*. Wired. https://www.wired.co.uk/article/hamburg-gdpr-johannes-caspar

Schelling, T. (1966). *Arms and Influence*. Yale University Press.

Schopenhauer, A. (1893). *Erystyka, czyli sztuka prowadzenia sporów* [*Eristics or the art of dispute*]. Księgarnia Teodora Paprockiego.

Schuman, T.D. (J. Bezmenow). (1984). *Love Letter to America*. W.I.N. Almanac Panorama.

Schwartau, W. (1996). *Information Warfare*. Thunder's Mouth Press.

Scire, S. (2021, September). *Publishers hope fact-checking can become a revenue stream*. Nieman Lab. https://www.niemanlab.org/2021/09/publishers-hope-fact-checking-can-become-a-revenue-stream-right-now-its-mostly-big-tech-who-is-buying/

Scott, M. (2021a, August, 24). Extremist content is flourishing on TikTok: Report more than 1,000 videos featured anti-Semitic and racist content over a one-month period. *Politico*. https://www.politico.eu/article/tiktok-extremist-content-white-supremacy/

Scott, M. (2021b, September 22). Ahead of German election, Telegram plays radicalizing role. *Politico.* https://www.politico.eu/article/german-telegram-election-misinformation/

Scott, M. (2021c, October 25). Facebook did little to moderate posts in the world's most violent countries. *Politico.* https://www.politico.eu/article/facebook-content-moderation-posts-wars-afghanistan-middle-east-arabic/

Seven commandments of fake news. (2018). EUvsDisinfo. Retrieved from https://www.euvsdisinfo.eu/seven-commandments-of-fake-news-new-york-times-exposes-kremlins-methods

Shabo, M.E. (2008). *Techniques of propaganda and persuasion.* Prestwick House.

Shane, T. (2021). *The psychology of misinformation: How to prevent it.* First Draft. https://firstdraftnews.org/articles/the-psychology-of-misinformation-how-to-prevent-it/

Shekhovtsov, A. (2021). *The Rise and fall of a Polish agent of the Kremlin influence: The case of Janusz Niedźwiecki.* European Platform for Democratic Elections. https://www.epde.org/en/news/details/the-rise-and-fall-of-a-polish-agent-of-the-kremlin-influence-the-case-of-janusz-niedzwiecki.html

Shu, K., et al. (Eds.). (2020). *Disinformation, Misinformation, and Fake News in Social Media: Emerging Research Challenges and Opportunities.* Springer Nature.

Skippage, R. (2021). *Is public service media doing enough to tackle misinformation?* Reuters Institute. https://reutersinstitute.politics.ox.ac.uk/public-service-media-doing-enough-tackle-misinformation

Smagliy, K. (2018). *Hybrid analytica: Pro-Kremlin expert propaganda in Moscow, Europe and the USA: Case study of think tanks and universities.* Institute of Modern Russia. https://www.underminers.info/publications/hybridanalytica

Smith, D. (2014). *Banned mind control techniques unleashed: Learn the dark secrets of hypnosis, manipulation, deception, persuasion, brainwashing and human psychology.* Make Profits Easy LLC.

Solsman, J. (2021, October 5). *YouTube says it's removed 1 million videos for COVID-19 misinformation.* CNET. https://www.cnet.com/news/youtube-says-its-removed-1-million-videos-for-covid-19-misinfo/

Soviet active measures in the West and the Developing World. (2020). Psywar.org. https://www.psywar.org/content/sovietActiveMeeasures

Spalding, R. (2022). *War without rules: China's playbook for global domination.* Sentinel.

Steffenhagen, M. (2021, August 20). *The Italian disinformation networks flying under Facebook's radar.* Coda. https://www.codastory.com/waronscience/italy-disinformation-facebook/

Stencel, M., & Luther, J. (2020). *Annual Census Finds Nearly 300 Fact-checking Projects Around the World: Growth is Fueled by Politics, Protests and Pandemic.* Duke Reporter's Lab. https://reporterslab.org/annual-census-finds-nearly-300-fact-checking-projects-around-the-world/

Stencel, M., & Luther, J. (2021). *Fact-checking census shows slower growth: The number of new projects dipped, even as fact-checking reached more countries than ever*. Duke Reporter's Lab. https://reporterslab.org/tag/fact-checking-census/

Stencel, M., Ryan, E., & Luther, J. (2022). *Fact-checkers extend their global reach with 391 outlets*. Duke Reporter's Lab, https://reporterslab.org/fact-checkers-extend-their-global-reach-with-391-outlets-but-growth-has-slowed/

Stokel-Walker, Ch. (2021a, July). Who is behind the online abuse of black England players and how can we stop it? *NewStatesman*. https://www.newstatesman.com/science-tech/social-media/2021/07/who-behind-online-abuse-black-england-players-and-how-can-we-stop

Stokel-Walker, Ch. (2021b). *Britain tamed Big Tech and nobody noticed*. Wired. https://www.wired.co.uk/article/age-appropriate-design-code-big-tech

Stopfake. (2022). *O nas* [*About us*]. https://www.stopfake.org/pl/o-nas-pl

Stradner, I., & Agrawal, P. (2021). *It is too easy to troll like a Russian*. Defense One. https://www.defenseone.com/ideas/2021/07/its-too-easy-troll-russian/183108/

Strategia Bezpieczeństw Narodowego RP [*National Security Strategy od Poland*]. (2020). Biuro Bezpieczeństwa Narodowego. https://www.bbn.gov.pl/ftp/dokumenty/Strategia_Bezpieczenstwa_Narodowego_RP_2020.pdf

Stróżyk, J. (2020). *Wybrane problemy międzynarodowej współpracy wywiadowczej. Czy NATO ma wywiad?* [*Selected problems of international intelligence cooperation. Does NATO have its intelligence service?*] Wydawnictwa Uniwersytetu Warszawskiego.

Stubbs, R. (2004). *Hearts and minds in guerrilla warfare: The Malayan Emergency 1948–1960*. Marshall Cavendish Intl.

Sumpter, D. (2018). *Outnumbered: From Facebook and Google to Fake News and filter-bubbles – The algorithms that control our lives*. Bloomsbury Sigma.

Sun Tzu. (2005). *Sztuka wojny* [*The art of war*]. Helion.

Sweet, J. (2021, July). Can Disinformation Be Stopped? Scholars' perspectives on a pervasive new threat. *Harvard Magazine*. https://www.harvardmagazine.com/2021/07/features-disinformation

Świerczek, M. (2018). *System matrioszek, czyli dezinformacja doskonała. Wstęp do zagadnienia* [The matrioshka system, or perfect disinformation. An introduction]. *Przegląd Bezpieczeństwa Wewnętrznego*, 19, 210–228.

Świerczek, M. (2020). *Największa klęska polskiego wywiadu. Sowiecka operacja dezinformacyjna „Trust" 1921–1927* [*Poland's greatest intelligence failure. The Soviet disinformation operation "Trust" 1921–1927*]. Fronda.

Szpyra, R. (2003). *Militarne operacje informacyjne* [*Military information operations*]. Akademia Obrony Narodowej.

Tackling disinformation: Information on the work of the EEAS Strategic Communication Division and its task forces. (2021). EEAS. https://eeas.europa.eu/headquarters/headquarters-homepage_en/

Tackling misinformation one algorithm at a time. (2021). Oxford Internet Institute. https://www.oii.ox.ac.uk/blog/tackling-misinformation-one-algorithm-at-a-time/

Talant, B. (2021). *How journalists can address misinformation on Telegram.* Reuters Institute. https://reutersinstitute.politics.ox.ac.uk/how-journalists-can-address-misinformation-telegram (Accessed October 13, 2021).

Taylor, J. (2021, July 24). Facebook forced to limit misinformation spread via WhatsApp amid Sydney lockdown. *The Guardian.* https://www.theguardian.com/australia-news/2021/jul/24/facebook-forced-to-limit-sydney-lockdown-misinformation-spread-via-whatsapp (Accessed August 16, 2021).

Ten simple ideas to regulate online political advertising in the UK. (2021). Who Targets Me. https://whotargets.me/en/ten-simple-ideas-to-regulate-online-political-advertising-in-the-uk/

Ten transatlantic principles for a healthy online information space: Endorse them here. (2021). GlobSec. https://www.globsec.org/publications/10-transatlantic-principles-for-a-healthy-online-information-space-endorse-them-here/

Tewari, S. (2021). *Swiss embassy urges Chinese media to remove articles about scientist.* BBC News. https://www.bbc.com/news/world-asia-china-58168588

The Digital Services Act package. (2021). European Commission. https://www.ec.europa.eu/digital-single-market/en/digital-services-act-package

The Global Disinformation Order: 2019 Global Inventory of Organised Social Media Manipulation. (2019). Oxford University. https://www.oii.ox.ac.uk/news-events/news/use-of-social-media-to-manipulate-public-opinion-now-a-global-problem-says-new-report/

The GRU's galaxy of Russian-speaking websites. (2022). Open Facto. https://openfacto.fr/2022/01/27/the-grus-galaxy-of-russian-speaking-websites/

The landscape of hybrid threats. A conceptual model. (2021). Hybrid Centre of Excellence. https://www.hybridcoe.fi/publications/the-landscape-of-hybrid-threats-a-conceptual-model/

The Media Literacy Index 2019: Just think about it. (2019). Open Society Institute. https://osis.bg/?p=3356&lang=en

The Nobel Peace Prize 2021. (2021). https://www.nobelprize.org/prizes/peace/2021/press-release/

The Overton Window. (2021). Mackinac Center. https://www.mackinac.org/OvertonWindow

Theohary, C. A. (2018). *Information warfare. Issues for Congress.* Congressional Reasearch Service.

Thomas, T. L. (2004). Russia's reflexive control theory and the military. *Journal of Slavic Military Studies*, 17(2), 237–256. http://dx.doi.org/10.1080/13518040490450529

Thomas, T. L. (2010). Russian information warfare theory: The consequences of August 2008. In S. J. Blank & R. Weitz (Eds.), *The Russian Military Today and*

Tomorrow: Essays in Memory of Mary Fizgerald* (pp. 265–300). Strategic Studies Institute.

Timmins, B. (2021, July 29). *Twitter works with news sites to tackle disinformation.* BBC. https://www.bbc.com/news/business-58065463

Treverton, G., Thvedt, A., Chen, A. R., Lee, K., & McCue, A. (2018). *Addressing hybrid threats.* Swedish Defence University.

Turton, W., & Jacobs, J. (2021, July 6). *Russia 'Cozy Bear' Breached GOP as Ransomware Attack Hit.* Bloomberg. https://www.bloomberg.com/news/articles/2021-07-06/russian-state-hackers-breached-republican-national-committee

Tversky, A., & Kahneman, D. (1973). *Judgment under uncertainty: Heuristics and Biases.* US Department of Commerce. https://www.apps.dtic.mil/sti/pdfs/AD0767426.pdf

Twetman, H., Pamment, J., Nothhaft, H., & Fjällhed, A. (2019). *The role of communicators in countering the malicious use of social media.* NATO Strategic Communications Centre of Excellence. https://stratcomcoe.org/publications/the-role-of-communicators-in-countering-the-malicious-use-of-social-media/101

Twitonomy. (2022). *Twitter #analytics and much more.* https://www.twitonomy.com/

Tworek, H., & Lee, Y. (2021, April 20). *Lessons from South Korea's approach to tackling disinformation.* Brookings. https://www.brookings.edu/techstream/lessons-from-south-koreas-approach-to-tackling-disinformation/

UK Government Communication Service. (2020). *RESIST Disinformation: A toolkit.* https://www.fundacioncarolina.es/wp-content/uploads/2020/11/Toolkit-UK.pdf

UK Government Communication Service. (2021). *RESIST 2: Counter Disinformation toolkit.* https://gcs.civilservice.gov.uk/publications/resist-2-counter-disinformation-toolkit/

UNESCO. (2003). *The Prague Declaration: Towards an Information Literate Society.* https://www.google.com/search?client=firefox-b-d&q=unesco+prague+deklaration+2003

Urbani, S. (2019, October). *Verifying online information.* First Draft. https://firstdraftnews.org/long-form-article/verifying-online-information/

Urbani, S. (2021, April). *An introduction to live audio social media and misinformation.* First Draft. https://firstdraftnews.org/articles/clubhouse-facebook-and-twitter-live-audio-and-misinformation/

Ustawa z dnia 26 stycznia 1984 r. Prawo prasowe [*Media law*]. (1984). Dz.U. z 1984 r. Nr 5, poz. 24. http://isap.sejm.gov.pl/isap.nsf/DocDetails.xsp?id=wdu19840050024

Ustawa z dnia 6 czerwca 1997 r. Kodeks karny [*Criminal Code*]. (1997). Dz.U. z 1997 r. Nr 88, poz. 553. https://isap.sejm.gov.pl/isap.nsf/download.xsp/WDU19970880553/U/D19970553Lj.pdf

Ustawa z dnia 5 sierpnia 2010 r. o ochronie informacji niejawnych [Protection of classified information]. (2010). Dz.U. z 2010 r. Nr 182, poz. 1228. https://isap.sejm.gov.pl/isap.nsf/download.xsp/WDU20101821228/U/D20101228Lj.pdf

US State Department. (1986). *Active Measures: A Report on the Substance and Process of Anti-U.S. Disinformation and Propaganda Campaigns*.https://insidethecoldwar.org/sites/default/files/documents/Soviet%20Active%20Measures%20Substance%20and%20Process%20of%20Anti-US%20Disinformation%20August%201986.pdf

Van Dongen, T. (2021). *Assessing the Threat of Covid 19-related Extremism in the West*. ICCT Publications. https://icct.nl/publication/assessing-the-threat-of-covid-19-related-extremism-in-the-west-2/

Víchová, V., & Janda, J. (2018). *The Prague Manual How to counter the Kremlin's influence in Europe*. Federal Academy for Security Policy. https://www.baks.bund.de/sites/baks010/files/working_paper_2018_22.pdf

Vienna Document 2011 on Confidence- and Security-Building Measures. (2011). OSCE. https://www.osce.org/files/f/documents/a/4/86597.pdf

Vincent, J. (2017, September 4). *Putin says the nation that leads in AI 'will be the ruler of the world'*. The Verge. https://www.theverge.com/2017/9/4/16251226/russia-ai-putin-rule-the-world

Volkoff, V. (2004). *La désinformation arme de guerre*. Editions l'Age d'Homme.

Volkoff, V. (1999). *Petite histoire de la désinformation. Du cheval de Troie à internet*. Éditions du Rocher.

Waddell, K. (2016, October). Internet blackouts can seriously damage a country's economy. *The Atlantic*. https://www.theatlantic.com/technology/archive/2016/10/the-global-economic-damage-of-internet-blackouts/503093/

Waltzman, R. (2017). *The Weaponization of information. The need for cognitive security. Testimony presented before the Senate Armed Services Committee, Subcommittee on Cybersecurity on April 27, 2017*. RAND Corporation. https://www.rand.org/content/dam/rand/pubs/testimonies/CT400/CT473/RAND_CT473.pdf

Warden, J. (1995). The enemy as a system. *Airpower Journal*, 9(1), 41–55. https://www.airuniversity.af.edu/Portals/10/ASPJ/journals/Volume-09_Issue-1-Se/1995_Vol9_No1.pdf

Wardle, C., & Derakhshan, H. (2017). *Information disorder: Toward an interdisciplinary framework for research and policymaking*. Council of Europe. https://www.firstdraftnews.org/latest/coe-report/

Watanabe, F. (1997). Fifteen axioms for intelligence analysts. *Studies in Intelligence*, 41(5), 45–47.

Weiss, M. (2020). *Aquarium leaks: Inside the GRU's psychological warfare program*. 4FreeRussia. https://www.4freerussia.org/aquarium-leaks-inside-the-gru-s-psychological-warfare-program/

Wenerski, Ł., & Kacewicz, M. (2017, April). *Russian soft power in Poland: The Kremlin and pro-Russian organizations*. Political Capital. https://www.politicalcapital.hu/pc-admin/source/documents/PC_NED_country_study_PL_20170428.pdf

What does it actually mean when a company says, "we do not sell your data"? (2021). The Markup. https://themarkup.org/ask-the-markup/2021/09/02/what-does-it-actually-mean-when-a-company-says-we-do-not-sell-your-data

What is Disinformation? (2018). Prevency. https://www.prevency.com/en/what-is-disinformation

What's driving journalism of the future? (2021, September 21). *HamburgNews*. https://hamburg-news.hamburg/en/congresses-events/whats-driving-journalism-future

WHOIS. (2022). *Blazing fast web hosting for your domain*. https://www.whois.com

Wildon, J., & Gildejeva, K. (2021). *Assessing The scale of German language disinformation communities on Telegram*. Logically. https://www.logically.ai/articles/german-language-disinformation-telegram

Wilk, A., & Żochowski, P. (2021). *The Zapad-2021 exercises. Russian strategy in practice*. Center for Eastern Studies. https://www.osw.waw.pl/en/publikacje/osw-commentary/2021-09-03/zapad-2021-exercises-russian-strategy-practice

Williams, H. J., & Blum, I. (2018). *Defining second generation open source intelligence (OSINT) for the defense enterprise*. RAND Corporation. https://www.rand.org/pubs/research_reports/RR1964.html

Williams, R. (1994, December 9). 'Guardian' journalist recruited by the KGB. *The Independent*. https://www.independent.co.uk/news/guardian-journalist-recruited-by-the-kgb-1386978.html

Winter, Ch. (2019). *Daesh propaganda, before and after its collapse*. NATO Strategic Communications Centre of Excellence. https://www.stratcomcoe.org/daesh-propaganda-and-after-its-collapse

Wirtschafter, V. (2021). *The challenge of detecting misinformation in podcasting*. Brookings. https://www.brookings.edu/techstream/the-challenge-of-detecting-misinformation-in-podcasting/

Wiszniewski, M. (1876). *Charaktery rozumów ludzkich* [*The nature of human reasoning*]. Wydawnictwo S. Lewentalla.

Wohlstetter, R. (1962). *Pearl Harbor: Warning and Decision*. Stanford University Press.

Wojnowski, M. (2015a). The concept of „modern warfare" formulated by strategists of the Russian Federation Armed Forces Headquarters. *Przegląd Bezpieczeństwa Wewnętrznego*, 13(7), 13–39.

Wojnowski, M. (2015b). Reflective management as a paradigm of the Russian information and psychological operations in 21st century. *Przegląd Bezpieczeństwa Wewnętrznego*, 12(7), 11–36.

Wojnowski, M. (2016). Alexander Dougin's concept of a net war as a tool towards realization of geopolitical goals of the Russian Federation. *Przegląd Bezpieczeństwa Wewnętrznego*, 16(9), 11–37.

Wojnowski, M. (2021). *Rosyjska ingerencja w amerykańskie wybory prezydenckie w latach 2016 i 2020 jako próba realizacji rewolucyjnego scenariusza walki informacyjnej*

[*Russian interference in the 2016 and 2020 US presidential elections as an attempt to implement a revolutionary information warfare scenario*]. Warsaw Institute. https://warsawinstitute.org/pl/rosyjska-ingerencja-w-amerykanskie-wybory-prezydenckie-w-latach-2016-2020-jako-proba-realizacji-rewolucyjnego-scenariusza-walki-informacyjnej/

Wolters, W., Stricklin, C., Carey, N., & McBride, M. K. (2021). *The psychology of (dis) information. A primer on key psychological mechanisms*. CNA. https://www.cna.org/reports/2021/10/psychology-of-disinformation-key-psychological-mechanisms

Woodhams, S., & Migliano, S. (2021). *Government internet shutdowns cost $5.5 billion in 2021*. TOP10VPN. https://www.top10vpn.com/research/cost-of-internet-shutdowns/

Woolley, S. (2020a). *The reality game*. Public Affairs.

Woolley, S. (2021b). *How can we stem the tide of digital propaganda?* Centre for International Governance Innovation. https://www.cigionline.org/articles/how-can-we-stem-the-tide-of-digital-propaganda/

Woolley, S. (2021c). *We need platforms that prioritize human rights and democracy over profit*. Centre for International Governance Innovation. https://www.cigionline.org/articles/we-need-platforms-that-prioritize-human-rights-and-democracy-over-profit/

Woolley, S., & Sawris, M. (2021). *Global democracies need to align to fight disinformation*. Wired. https://www.wired.com/story/opinion-global-democracies-need-to-align-to-fight-disinformation/

Woszczyńska, A. (2011). W kręgu teorii kłamstwa. *Językoznawstwo*, 5(1), 191–197. https://www.bazhum.muzhp.pl/media/files/Jezykoznawstwo_wspolczesne_badania_problemy_i_analizy_jezykoznawcze-r2011-t5-s191-197.pdf

Wuhan lab leak theory: How Fort Detrick became a centre for Chinese conspiracies. (2021, August 13). BBC News. https://www.bbc.com/news/world-us-canada-58273322

Yaffa, J. (2014, July 14). Dmitri Kiselev is redefinig the art of Russian Propaganda. *New Republic*. https://newrepublic.com/article/118886/dmitri-kiselev-redefining-art-russian-propaganda

Zamecnik, A. (2021, September 9). *An army of volunteer 'elves' fights disinfo in the Czech Republic*. Coda. https://www.codastory.com/disinformation/volunteers-fight-disinfo-czech-republic/

Zimbardo, P., & Gerrig, R. J. (2009). *Psychology and Life*. ALLYN & BACON.

Zurkowski, P. (1974). *The information services environment. Relationships and priorities*. National Commission on Libraries. https://www.files.eric.ed.gov/fulltext/ED100391.pdf

Index

Printed in the United States
by Baker & Taylor Publisher Services